Learning Materials in Biosciences

Learning Materials in Biosciences textbooks compactly and concisely discuss a specific biological, bio-medical, biochemical, bioengineering or cell biologic topic. The textbooks in this series are based on lec-tures for upper-level undergraduates, master's and graduate students, presented and written by authoritative figures in the field at leading universities around the globe.

The titles are organized to guide the reader to a deeper understanding of the concepts covered.

Each textbook provides readers with fundamental insights into the subject and prepares them to indepen-dently pursue further thinking and research on the topic. Colored figures, step-by-step protocols and take-home messages offer an accessible approach to learning and understanding.

In addition to being designed to benefit students, Learning Materials textbooks represent a valuable tool for lecturers and teachers, helping them to prepare their own respective coursework.

More information about this series at http://www.springer.com/series/15430

Narine Sarvazyan

Editor

Tissue Engineering

Principles, Protocols, and Practical Exercises

 Springer

Editor
Narine Sarvazyan
The George Washington University
Washington, DC
USA

ISSN 2509-6125   ISSN 2509-6133   (electronic)
Learning Materials in Biosciences
ISBN 978-3-030-39697-8   ISBN 978-3-030-39698-5   (eBook)
https://doi.org/10.1007/978-3-030-39698-5

This Springer imprint is published by the registered company Springer Nature Switzerland AG
The registered company address is: Gewerbestrasse 11, 6330 Cham, Switzerland

Introduction

The goals of this textbook Tissue engineering and regenerative medicine (TERM) is a newly emerged interdisciplinary branch of science with high clinical relevance. Here, we attempted to overview the main TERM principles followed by simple demos and student exercises – all done using minimal specialized equipment and supplies. As such, this book can serve as a suitable educational material for anyone who wants to conduct a practical course introducing students to this complex and rapidly advancing field of science. We intentionally avoided exercises that rely on specialized equipment such as electrospinning, magnetic bead sorter, microfluidic chamber, or 3D printer. In contrast, described protocols can be conducted using common lab equipment, and the didactic part of the book can be easily read by students from all backgrounds including high school or non-biology major students. Moreover, since not all educational institutions might have a functional vivarium, we also included suggestions for alternative experiments that do not require certified animal facilities and approved animal protocols. Instead, instructors and students can rely on tissue sources from a local abattoir, farmer's market, or using commercial cell lines. By completing exercises described in this book, students will become familiar with (i) the basic principles of disassembling and reassembling tissues, (ii) key TERM terminologies, and (iii) methods of analyzing engineered tissues. Students will also gain practical experience in using online resources and literature searches to extract relevant TERM protocols and to perform the key steps of the procedures.

It is important to emphasize that by no means this book attempts to cover the advances and complexities of the TERM field. For this, interested readers are referred to many excellent comprehensive books or reviews that encompass different aspects of tissue engineering.

About the author Narine Sarvazyan, Ph.D., Editor of this book, is a Professor of Pharmacology and Physiology at the George Washington University School of Medicine and Health Sciences. Her lab focuses on exploring the mechanisms of cardiac arrhythmias, cardiotoxicity of cancer drugs, the role of environmental contaminants in heart disease, and tissue engineering-based therapies for heart and vessel repair. In the Fall of 2017, she received the Fulbright US Scholar Award that enabled her to develop and conduct an introductory, hands-on course in tissue engineering, which was held on the premises of the Orbeli Institute of Physiology, Armenian National Academy of Sciences, using the facilities and equipment of the Tissue Engineering and Immunology Laboratory led by Dr. Zaruhi Karabekian. Dr. Karabekian served as a Co-instructor in the course, while two of her lab members, Drs. Hovhannes Arestakesyan and Vahan Grigoryan, provided valuable contributions during and after the course as teaching assistants. This book is an outcome of this course. Seven out of 12 students who took the course contributed to the specific chapters of this book, which is reflected on the content page. Hasmik Mikaelyan, Regional Coordinator of the Fulbright programs in Armenia, and Dr. Naira Ayvazyan, Director of the Orbeli Institute of Physiology, are gratefully acknowledged for their support and encouragement throughout these activities.

Pre-course organization   Below is a brief outline of the administrative steps that were taken to organize the abovementioned TERM course in Yerevan, Armenia. Albeit basic, this information might be useful for anyone organizing similar educational activities.

- An online advertisement using social media channels and Instapage sign-up was used to inform interested students and the general public. Instapage is a paid service, but a similar sign-up page can be easily established for free. The main goal of these wide advertisement efforts was to recruit an interdisciplinary group of students from any STEM fields as well as of different career stages: from high school students to recent PhDs.
- A Doodle poll was created to schedule individual student interviews. Applicants were also asked to submit, in advance, their CV and a short letter of interest. During the follow-up interview, the student's level of interest and his/her time commitment were evaluated.
- Students were informed about strict attendance rules and a requirement to commit to a substantial number of post-session hours to complete homework assignments and to perform their own experiments.
- Participants were split into small interdisciplinary teams. Team assignment was done by the course director with the goal of forming maximally diverse teams. This can be also done randomly or by a draw.
- A sign-up sheet was used to record student attendance throughout the course. The use of cell phones in class was forbidden unless noted otherwise by the instructor. Each student was allowed one unexcused absence throughout the semester. The day of the absence had to be coordinated with the course director and team members. Each additional missed class, regardless of the underlying reason, led to a student losing points from his/her overall grade.
- Three-hour-long, in-class mandatory sessions were held twice a week. Laboratory was available for any additional hours needed to perform team assignments between the sessions. At the end of the course, each team prepared a poster and an oral presentation for the public.

Minimal required equipment   The list of the major equipment required to perform described protocols includes:

- Laminar hood or safety cabinet suitable for cell culture
- Cell culture incubator with CO_2 tank
- Inverted phase microscope
- Fluorescent microscope
- Cell centrifuge
- Water bath with shaker
- Refrigerator with freezer
- pH meter, magnetic mixer, water distiller, UV sterilizer
- Liquid nitrogen tank

List of required supplies includes plasticware; surgical instruments; glassware; cell culture media; filter units; bovine serum; basic salts; viability assay reagents; fluorescent probes; histology dyes; growth factors; cell attachment factors such as fibronectin, laminin, and gelatin; and other noted common reagents. Additional highly desirable items include 80 °C freezer and confocal microscope. For experiments involving rodents, a functional vivarium and an approved animal protocol for extracting animal tissues are required. In the absence of those, key experiments can be performed using anesthetized live fish or freshly excised animal tissue from a local abattoir. The table illustrating how animal tissue can be used in the most efficient way through the course is included. In the absence of commercial cell lines, primary cell cultures obtained during week 4 can be used for viability and cell plating experiments.

Suggested Course Outline

This textbook is designed for a one-semester course with two 3-hour mandatory sessions held each week. During the first hour of the first session, teams present results of their homework from the previous week followed by group discussion. During the second hour, the instructor goes over the material discussed in the assigned chapter and answers any questions students might have about it. After a short break, the instructor conducts a demonstration that covers the basic steps of that week's experiment. Before the second session, teams find and read relevant protocols from the literature needed to conduct experiments of their own. During the second session, the instructor plays only an advisory role, while students work in teams using the information they learned during the first session and what they extracted from the literature search. Homework consists of compiling outcomes of the team experiments into three to six slide presentations. The exercise complexity increases as students learn how to work with tissues and use available lab equipment.

Described in-class exercises are designed to be completed within 3 hours of the second session. However, students are welcome to use the lab for any additional time needed to troubleshoot or complete these experiments. Homework assignments involve efforts of all team members and should be completed within 2–3 days after each session. The book intentionally includes only the outlines of the sample protocols since at the very beginning of the course students are taught how to search available online portals. Performing these targeted searches and comparing detailed protocols from different labs help students develop critical thinking skills as to what is important and what can be omitted or modified.

It is advised to recruit at least one teaching assistant who can help to supervise students during experiments isolate cells for demos and team exercises, prepare and change cell culture media, deliver tissue from an abattoir, clean up after tissue dissection, and help with other basic activities.

During weeks 12 and 13, the teams are tasked to create the simplest version of engineered tissue of their choice. This is arguably the most involved and exciting part of the course. During this time, team members work based on their own schedule and coordinate experiments between themselves, while the instructor and teaching assistant are available for any questions. Oral and poster presentations are then prepared based on the students' hands-on experience.

Below is a suggested outline for a one-semester six-credit-hour course. Its simplified form is shown in ◘ Table 1. The course can be easily modified for 3-credit-hour course that spans two semesters.

- **Week 1: Introduction – Reference Search, Image, and Data Analysis**

Session I

1st hour. Overview of TERM field followed by Q&A.

2nd hour. Students introduce themselves to the group. Teams are formed and chose their team color. The instructor shows major lab equipment to be used in the course and goes over relevant safety rules.

3rd hour. Demo: The instructor demonstrates how to extract relevant protocols from online open sources and uses sample images to show the main features of ImageJ and how to use Mendeley software for references.

◻ Table 1 The suggested flow of subjects and exercises for a one-semester six credit hour course. White arrows – tissues should be frozen, blue arrows – cells should be cultured

Chapter/week	Symbol	Main procedure	Tissue source in bold (anesthetized animals from vivarium)	Tissue source in bold (tissue from abattoir or market)
CH1. Data analysis		Quantitative analysis of online images	No need for animal tissue. Use of online images	
CH2. Organ structure		Dissection & organ cannulation	Adult rat *(freeze tails & cannulated organs)*	Chicken or fish *(freeze skins & cannulated organs)*
CH3. ECM & adhesion molecules		Collagen extraction, coverslips preparation	Stored rat tails	Pig or chicken skin, cow or pig tendons, fish scales
CH4. Cell isolation		Cell isolation by digestion, followed by plating	Neonatal rats: isolation of cardiac myocytes & fibroblasts	Bovine, porcine, ovine aorta or joints
CH5. Functional assays		Dynamic & static cell assays	Myocytes from week 4	Cells from week 4
CH6. Cell culture		Media preparation, cell splitting & storage	Fibroblasts from week 4	Cells from week 4
CH7. Imaging		Staining with multiple dyes	Fibroblasts from week 6	Cells from week 6
CH8. Stem cells		Isolation of adipose stem cells, immunostaining	Adult rat adipose tissue	Chicken, pig or cow adipose tissue
CH9. Scaffolds		Perfusion with detergents	Organs stored from week 3 or from another **adult rat**	Organs stored from week 3 or from fresh **fish or chicken**
CH10. Casting & 3D printing		Making molds & casting	No need for animal tissue. Use of biocompatible materials	
CH11. Bioreactors		Making a bioreactor	No need for animal tissue. Use of plastic and tubing	
CH12. Presenting data		Posters and oral presentations based on team experiments	Available & instructor-approved animal sources to be used for team experiments	

Homework: Students are asked to download Mendeley and ImageJ software and familiarize themselves with their main features. Reading assignment: review TERM field history (Kaul & Ventikos, TISSUE ENGINEERING B, v.21, N.2, 2015 or its more recent equivalent).

Session II

Each team selects a pair of sample images to be quantitatively compared. Any free online publication or other open sources can be used to obtain two images. They can include cell cultures, decellularized tissues, scaffolds, or any other TERM-relevant material. Teams are then tasked with extracting three different quantitative measurements from selected pairs of images using ImageJ tools.

Homework: Teams are asked to compile a brief PowerPoint presentation illustrating the results of image analysis by each team.

▪ Week 2: Organ Structure and Vascularization

Session I

1st hour. Homework presentations followed by group discussion.

2nd hour. Lecture based on ▶ Chap. 2, followed by Q&A.

3rd hour. Demo: The instructor demonstrates how to make perfusion cannula, heparinize and anesthetize the animal, perform basic dissection, and cannulate either the heart, liver, or lungs.

Homework: Teams are asked to decide on the organ they want to cannulate and read about its function, anatomy, and vascular supply.

Session II

Using tape of chosen color, each team labels its tools, plasticware, and reagents. Teams make their own cannulas and perfusion solutions and excise and cannulate the excised organ of their choice. Notes and images of the entire process are taken through the experiment. After tissue becomes blanched upon blood removal, can-nulated organs are placed in a freezer to be used later in the course.

Homework: Teams are tasked to create a detailed protocol to describe performed procedures that include images taken at every step of their cannulation experiments.

■ Week 3: Extracellular Matrix and Adhesion Molecules

Session I

1st hour. Homework presentations followed by group discussion.

2nd hour. Lecture based on ▶ Chap. 3, followed by Q&A.

3rd hour. Demo: The instructor shows the main steps involved in the extraction of rat tail collagen and how to treat glass coverslips, plastic plates, and scaffold materi-als with different adhesion proteins.

Homework: Team members are instructed to find and read a recent review article on cell adhesion molecules and their roles in creating engineered tissues.

Session II

Students execute the first part of the collagen isolation protocol. In addition, each team covers cell culture and glass coverslips with solutions of gelatin, collagen, fibro-nectin, or albumin and label and store them for next week's experiment.

Homework: Team members take turns during the next few days to complete the collagen isolation protocol and to document all the steps involved.

■ Week 4: Isolating Cells from Tissue

Session I

1st hour. Homework presentations followed by discussion.

2nd hour. Lecture based on ▶ Chap. 4, followed by Q&A.

3rd hour. Demo: The instructor shows how to isolate neonatal rat ventricular myocytes and fibroblasts using trypsin-based digestion of manually dispersed tissue. Alternatively, isolation of chondrocytes from cartilage or endothelial cells from the aorta is shown. Freshly excised porcine, bovine, or sheep tissues brought to the lab on ice from a local abattoir can be used for the latter procedures. The concept of perfu-sion-based digestion of cannulated organs using collagenase is also described.

Homework: Teams are tasked with searching the literature to find the simplest and most suitable detailed protocol to isolate cells from their organ of choice.

Session II

Based on their homework assignment, each team attempts to isolate cells based on either dispersion- or perfusion-based protocol followed by collagenase or trypsin digestion, centrifugation, and cell count. Isolated cells are then plated on coverslips prepared by the students during the previous session.

Homework: Teams are tasked with a detailed description of performed cell isolation steps, including phase-contrast images of the cells and calculation of cell yield and via-bility.

■ Week 5: Functional Assays and Toxicity Screening

Session I

1st hour. Homework presentations followed by discussion.

2nd hour. Lecture based on ▶ Chap. 5, followed by Q&A.

3rd hour. Demo: The instructor exposes available cells to different concentrations of toxins or oxidants followed by resazurin, LDH, or MTT assay. Assessment of cell contractions and other tissue-specific functional assays for toxicity assessment is also demonstrated and discussed. For example, the use of Fluo-4 to monitor calcium transients using plated cardiomyocytes from the previous week can be shown.

Homework: Based on a literature search, teams design an experiment in which passaged cells from the previous week are exposed to chemical or physical stress followed by cell viability assessment.

Session II

Teams test chosen stress conditions and methods to determine cell viability afterward. The goal is to create a viability curve that should have at least three-point control (i.e., no damage), midpoint (some damage), and max damage (i.e., all cell dead) points.

Homework: Teams use their experimental data to create a graph illustrating the effect of their chosen treatment on cell viability. Positive, negative, and blank samples must be included.

■ Week 6: Culturing Cells in 2D and 3D

Session I

1st hour. Homework presentations followed by group discussion.

2nd hour. Lecture based on ► Chap. 6, followed by Q&A.

3rd hour. Demo: The instructor demonstrates media preparation, passaging, counting, and storing cells using commercial cell lines. Alternatively, fibroblast or endothelial cells obtained during the week 4 demo can be kept in culture and used for these experiments. The concept of 3D culture by making spheroids using a hanging drop technique can be also shown and discussed.

Homework: Each team selects from literature three different protocols that can be used to culture cells from their organ of choice. A table comparing major differences and similarities in these three culturing conditions is then compiled.

Session II

Each team is given one plate of cells to passage using their own medium. The latter has to be prepared from media powder to which appropriate amounts of bicarbonate have to be added, followed by filtering and addition of serum and antibiotics. Treated coverslips from week 3 can be used to plate excess of the cells on different adhesion surfaces.

Homework: Each team prepares a detailed protocol that describes steps involved in making culture media, passaging the cells and monitoring their proliferation on different adhesion surfaces.

■ Week 7: Imaging, Staining, and Markers

Session I

1st hour. Homework presentations followed by discussion.

2nd hour. Lecture based on ► Chap. 7, followed by Q&A.

3rd hour. Demo: The instructor shows how to use different light paths on available fluorescent scopes to perform imaging of samples stained with different dyes. If a confocal setup is available, the use of Z-stacks to image tissue slices is also shown. Lastly, students are shown how to fix and stain cultured cells on a glass coverslip using multiple organelle-specific dyes.

Homework: Students familiarize themselves with the spectral properties of a given list of dyes to be used during Session II.

Session II

Using known dye spectra and list of available probes, each team selects the best dye combinations and filter settings to perform triple staining using DAPI, phalloidin, mitochondrial, and/or other organelle-specific markers.

Homework: Students are tasked with taking images of their stained samples and making a PowerPoint presentation that shows these images.

- **Week 8: Stem Cells and Basics of Immunology**

Session I

1st hour. Homework presentations followed by discussion.

2nd hour. Lecture based on ▸ Chap. 10, followed by Q&A.

3rd hour. Demo: The instructor shows steps involved in the isolation of adipose mesenchymal stem cells from adipose tissue and immunostaining for stem cell markers.

Homework: Students search the literature for the most suitable antibodies and protocols to stain stem cells from the demo session. List of available primary and secondary antibodies are given to students to narrow their choices.

Session II

Students stain adipose explant stem cell cultures with stemness markers of their choice, followed by imaging using a fluorescent microscope.

Homework: Students are tasked with making a PowerPoint presentation that includes images of stained cell cultures with and without primary antibodies.

- **Week 9: Scaffolds and Tissue Decellularization**

Session I

1st hour. Homework presentations followed by discussion.

2nd hour. Lecture based on ▸ Chap. 11, followed by Q&A.

3rd hour. Demo: The instructor shows an example of a decellularization protocol.

Homework: Teams are tasked with searching the literature to find the simplest and most suitable protocol to decellularize their organ of choice and how to record the endpoints of the decellularization process.

Session II

Teams perform whole-organ decellularization using their selected protocol. Fish or rats are recommended species, but frozen organs from larger animals can be also considered.

Homework: Team members take turns during the next few days to finish the decellularization process and to record its endpoints. PowerPoint slides showing tissue micro- and macro-appearance have to be prepared.

- **Week 10: Casting and 3D Bioprinting**

Session I

1st hour. Homework presentations followed by discussion.

2nd hour. Lecture based on ▸ Chap. 12, followed by Q&A.

3rd hour. Demo: The instructor shows how to create molds and illustrates differences between the physical properties of various scaffold materials.

Homework: Teams are tasked with searching the literature on how to make 3D models of their organ of choice using biocompatible materials.

Session II
Students make molds and cast different biocompatible materials to create a shape of their choice.
Homework: Teams prepare PowerPoint slide presentation summarizing the casting process.

■ **Week 11: Bioreactors**

Session I
1st hour. Homework presentations followed by discussion.
2nd hour. Lecture based on ► Chap. 13, followed by Q&A.
3rd hour. Demo: The instructor builds a simple bioreactor using available tools demonstrating its most important features.
Homework: Teams are tasked with searching the literature to find a design of feasible bioreactor suitable for long-term culturing of their organ of choice.

Session II
Each team builds a different type of bioreactor using available tools. The type of bioreactor can match the tissue that the team is planning to cultivate. This exercise will help students to build a more elaborated device for the actual recellularization during the following weeks.
Homework: Team members meet to decide how they will proceed with engineering an organ of their choice within the next 2 weeks. Methods, hypotheses to be tested, animal species from which obtain organs and cells, required reagents, assignment of individual tasks – all these items must be debated and presented to the instructor for further discussion.

■ **Weeks 12–13: Engineering Tissue of Choice**
Each team meets with the instructor individually to discuss the design of their planned experiments. For the next 2 weeks, teams work at their own established schedule to solve problems, collect data, and prepare draft presentations.

■ **Week 14: Presentation Workshop**

Session I
Lecture based on ► Chap. 12, discussion of draft *poster* presentations using instructor feedback, and finishing data collection.

Session II
Discussion of draft *oral* presentations using instructor feedback and finishing data collection
Homework: Team members finish their oral presentations and posters based on instructor feedback. Each student prepares a 1- to 2-page report on lessons learned, what can be improved and what worked particularly well throughout the course.

■ **Week 15: Final Team Presentations**

Session I
Poster and oral presentations by the teams for peer-to-peer assessment.

Session II
Final presentations to the rest of the class, family members, and interested public. Awards ceremony for the best poster and best oral presentation.

Preface

What is tissue engineering?   Tissue engineering and regenerative medicine (TERM) is a new interdisciplinary branch of science that combines knowledge from numerous scientific fields including biology, biochemistry, physics, chemistry, applied engineering, and medicine. TERM aims to restore damaged parts of the human body by rebuilding them using individual building blocks of biological tissues such as cells and the extracellular matrix that surrounds them. Most of the TERM approaches are based on a sequence of the following main steps depicted in ◘ Fig. 1. This includes extraction of patient's own cells, their amplification, differentiation to a specific phenotype, cell seeding into a biocompatible scaffold, and growing engineered tissues in vitro followed by their transplantation back to a patient. Other TERM approaches include stimulating damaged organs to repair themselves either by promoting endogenous cell production from resi-

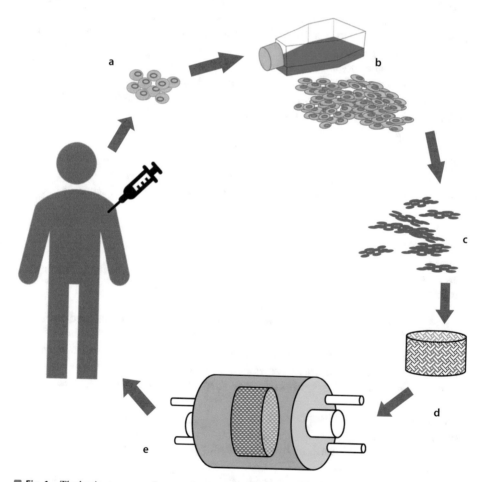

◘ **Fig. 1**   The basic sequence of steps to create engineered tissue: **a** sample collection, **b** cell isolation, **c** cell amplification and differentiation, **d** cell seeding in scaffolds, **e** culturing of cell-seeded scaffolds followed by implantation

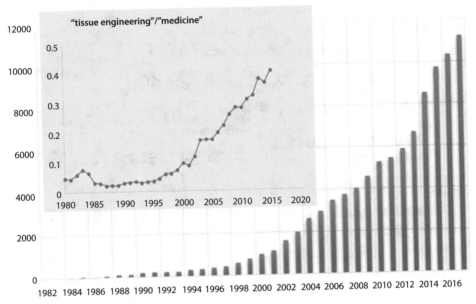

Fig. 2 The number of articles on PubMed that appears when searching for words 'tissue engineering'. The main graph shows the absolute numbers of papers, while the pink shaded insert expresses the same numbers as a ratio to a search word 'medicine'

dent stem cells or by populating implanted acellular scaffolds using the patient's own redifferentiated cells.

In the last two decades, the TERM field experienced exponential growth (Fig. 2). Today, it continues to expand both in scope and in the amount of involved research and translational efforts. Research articles relevant to TERM can be found in journals covering a multitude of disciplines (Fig. 3). Several TERM products, particularly skin substitutes, are already being used to treat patients, and many more are undergoing clinical trials. Examples of diseases that TERM ultimately aims to cure include osteoporosis, diabetes, heart disease, skeletomuscular defects, spinal injuries, kidney failure, venous deficiency, and many other deadly or previously untreatable diseases.

The initial long-term goal of TERM was to replace a damaged organ with a new, stronger one created from a patient's own cells – as depicted in a cartoon form in Fig. 4. Indeed, as of today, proof-of-principle simple forms of several tissues such as the skin, cartilage, vessels, and bone have been already created. Efforts to create more complex tissues and organs, such as tissue-engineered heart, kidneys, or lungs are ongoing. In addition to research and development efforts to recreate new organs suitable for transplantation, the TERM field has branched into several complementary directions. One of them is a new global industry to test drugs using small pieces of tissues made from a patient's own cells. The latter is called the *organ-on-a-chip* or, in the case of multiple interconnected engineered tissues, the *human-on-a-chip* field. Stem cell-based therapies, 3D bioprinting, and implantation of acellular scaffolds are other branches of rapidly expanding TERM field. These additional directions are depicted in Fig. 5 and will be discussed in more detail in the following chapters. Other emerging branches of tissue engineering include the use of injectable hydrogels for controllable

Environmental Engineering (3,866)   Software (1,674)   Cardiology and Cardiovascular Medicine (3,858)

Computer Science Applications (5,220)   Organic Chemistry (4,344)   Physiology (medical) (2,049)

Materials Chemistry (4,347)   General Materials Science (13,098)   Cancer Research (2,778)

Rheumatology (3,202)

Plant Science (3,532)   General Medicine (29,409)   Rehabilitation (3,061)

Virology (934)   Surgery (5,867)   Transplantation (3,502)

General Chemistry (8,662)   Biomaterials (36,405)

Histology (1,673)   Metals and Alloys (4,518)

Medicine (miscellaneous) (9,699)

Mechanical Engineering (8,310)   Bioengineering (26,855)

Pharmacology (3,432)

Multidisciplinary (3,381)   Biophysics (20,629)   Biotechnology (17,844)

Polymers and Plastics (4,211)   Modelling and Simulation (2,174)   Biochemistry (16,573)

General Engineering (7,804)   Molecular Biology (13,311)

Oncology (3,233)   Cell Biology (12,478)   Molecular Medicine (5,399)

Mechanics of Materials (10,660)   Physiology (3,333)

General Neuroscience (3,715)   Radiology Nuclear Medicine and imaging (6,212)   Genetics (6,934)

General Immunology and Microbiology (3,165)   Pollution (3,168)   Immunology (2,101)

Drug Discovery (2,612)   Developmental Biology (2,870)   Environmental Chemistry (3,110)

◘ Fig. 3 Word cloud chart from lens.org site that displays the top journal subjects while searching for "tissue engineering"

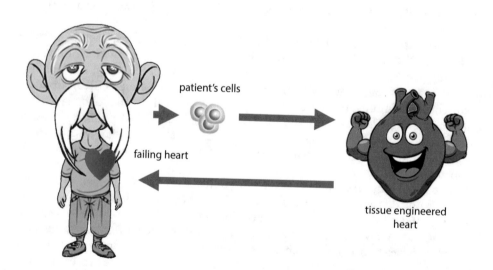

patient's cells

failing heart

tissue engineered heart

◘ Fig. 4 The long-term goal of the TERM field is to create replacement organs from a patient's own cells

drug delivery and the use of tissue engineering tools to create 3D tumor models. This is because a great amount of evidence suggests that response to anticancer drugs is very different when cells are grown in 3D vs 2D environment. Therefore, creating in vitro 3D models of tumor growth using tissue engineering approaches can greatly speed up testing and development of new anticancer treatments.

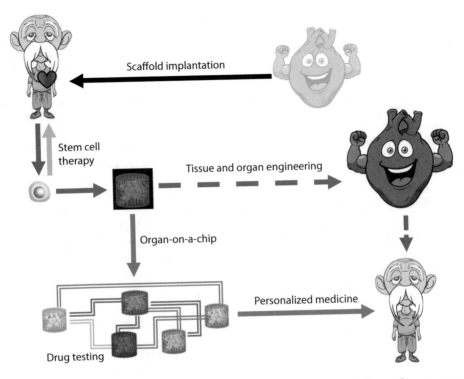

◘ Fig. 5 Current branches of the TERM field. In addition to the main goal of recreating new organs (red arrows), TERM sub-fields now include organ/human-on-a-chip field (blue arrow) for drug testing, implantation of decellularized scaffolds (black arrow), and stem cell therapy (green arrow)

A brief history of the TERM field TERM confirms that many current technologies are realized myths and dreams of the past. This list includes the ability of humans to fly, reading human thoughts, connecting via long distances, seeing far away objects, and other concepts that can be found in fairy tales of numerous cultures. In the case of the TERM field, many folk stories tell us about recreating human beings from drops of blood or other body fluids, magically growing parts of damaged organs or creating living creatures using in vitro-like conditions. In practical terms, animal skin, vessel or soft tissue grafts, and artificial materials to replace damaged human organs have been tried for thousands of years as documented by archeological evidence from Egyptian and Mayan civilizations. It continued throughout the Middle Ages with many of these procedures documented in various medieval paintings and manuscripts. The Era of Enlightenment brought more rigorous, evidence-based discoveries that demystified the main functions of the human body. The invention of the microscope led to direct observation of tissue structure, giving birth to cell theory that was followed by a much better understanding of organ functions and the development of physiology as a separate branch of science. In the early nineteenth century, Henry Levert did the first systematical in vivo evaluation of biocompatibility of different implant materials by testing platinum, gold, silver, and lead sutures. A century later, Alexis Carrel developed cell culture techniques, followed by the invention of organ perfusion apparatus. During World War II, several materials capable of osseointegration were discovered, enabling permanent anchorage of metal parts to the human skeleton. Judah Folkman was the next key figure in the development of the TERM field. His many contributions include showing the effects of various substrates on cell differentiation and proliferation rates. Students are encouraged to read more about discoveries leading to the current phase of the TERM field in a review of its historical origins by Kaul and Ventikos [1].

In its modern form, "tissue engineering" started to appear in the scientific literature in the mid-80s of the twentieth century. Its use and origins have been attributed to various authors. Initially, it was described as a field of science that aims to create artificial parts that could replace human organs or tissues or induce regeneration. Its definition was further defined by Joseph Vacanti and Robert Langer in their 1993 Science article [2]. To the general public term, "tissue engineering" became known when BBC made a report about "auriculosaurus" mouse with a human ear on its back. It was created by Charles Vacanti's lab in 1997 using bovine cartilage cells [3]. Since then, publications and interest in TERM-related fields skyrocketed, and soon after Tissue Engineering and Regenerative Medicine International Society was formed. Game-changing discovery by Shinya Yamanaka's lab [4] showing that it is possible to derive pluripotent stem cells from mature differentiated cells further fueled the growth of this exciting new branch of applied science.

How far are we from creating tissue-engineered organs?   The first whole organ to be reportedly bioengineered ex vivo and then implanted into the human body was the urinary bladder. Biopsies of such tissue-engineered bladder performed postoperatively revealed a proper three-layered structure, and no adverse events were observed [5]. However, several follow-up studies have revealed significant shortcomings of similarly created bladders. As an example, one can cite a more recent study [6] that reported no improvement in tissue-engineered bladder compliance or capacity, while serious adverse events surpassed acceptable safety standards. Similar scenarios occurred with other engineered tissues, where the initial proof-of-the-principle studies appeared to be highly successful only to be followed by comprehensive papers that revealed multiple problems and adverse clinical effects. Today, tissue engineering of relatively thin multilayered organs, such as the bladder, trachea, and skin, as well as larger and more complex organs, such as the heart, liver, or lungs, remains an active area of research. Ongoing efforts include enlarging the size of tissue constructs, improving their blood supply, increasing cell density, and tuning their mechanical properties. By addressing, in a systematic and consistent manner, these challenges, the field will be able to deliver on its original promise to cure diseased organs and tissues.

Narine Sarvazyan
Washington, DC, USA

References

1. H. Kaul, Y. Ventikos, On the genealogy of tissue engineering and regenerative medicine. Tissue Eng. Part B Rev. **21**(2), 203–217 (2015)
2. R. Langer, J.P. Vacanti, Tissue engineering. Science **260**(5110), 920–926 (1993)
3. Y. Cao, J.P. Vacanti, K.T. Paige, J. Upton, C.A. Vacanti, Transplantation of chondrocytes utilizing a polymer-cell construct to produce tissue-engineered cartilage in the shape of a human ear. Plast. Reconstr. Surg. **100**(2), 297–302; discussion 303–4 (1997)

4. K. Takahashi, S. Yamanaka, Induction of pluripotent stem cells from mouse embryonic and adult fibroblast cultures by defined factors. Cell **126**(4), 663–676 (2006)
5. A. Atala, S.B. Bauer, S. Soker, J.J. Yoo, A.B. Retik, Tissue-engineered autologous bladders for patients needing cystoplasty. Lancet **367**(9518), 1241–1246 (2006)
6. D.B. Joseph, J.G. Borer, R.E. De Filippo, S.J. Hodges, G.A. McLorie, Autologous cell seeded biodegradable scaffold for augmentation cystoplasty: Phase II study in children and adolescents with spina bifida. J. Urol. **191**(5), 1389–1395 (2014)

Contents

Contributors

Hovhannes Arestakesyan L. Orbeli Institute of Physiology, Yerevan, Armenia

Vardan Avetisyan Yerevan State University, Yerevan, Armenia

Vahan Grigoryan L. Orbeli Institute of Physiology, Yerevan, Armenia

Mariam Grigoryan Yerevan State Medical University, Yerevan, Armenia

Liana Hayrapetyan Yerevan State Medical University, Yerevan, Armenia

Zaruhi Karabekian L. Orbeli Institute of Physiology, Yerevan, Armenia

Astghik Karapetyan G.S.Davtyan Institute of Hydroponics Problems, Yerevan, Armenia

Artem Oganesyan Yerevan State Medical University, Yerevan, Armenia

Narine Sarvazyan The George Washington University, Washington, DC, USA

Arman Simonyan American University of Armenia, Yerevan, Armenia

Anna Simonyan Yerevan State University, Yerevan, Armenia

Abbreviations

AAD	Aminoactinomycin D	**H&E**	Hematoxylin and eosin stain
AA-PBS	Antibiotic-antimycotic phosphate-buffer saline	**HEK cells**	Human embryonic kidney cells
Ad-MSC	Adipose tissue-derived mesenchymal stem cells	**HEPA**	High-efficiency particulate absorber
AM	Acetoxymethyl	**HEPES**	(4-(2-hydroxyethyl)-1-piperazineethanesulfonic acid)
bFGF	Basic fibroblast growth factor		
		ICAM	Intercellular cell adhesion molecules
CAMs	Cell adhesion molecules		
C-CAM	Cell-cell adhesion molecules	**IgSF**	Immunoglobulin superfamily
CHAPS	3-[(3-cholamidopropyl) dimethylammonio]-1-propane-sulfonate	**IMDM**	Basic DMEM modified by Iscove
		iPSC	Induced pluripotent stem cells
CTL	Cytotoxic T lymphocytes		
		LDH	Lactate dehydrogenase
DAPI	6-diamidino-2-phenylindole		
DCT	Decellularized tissue	**MEDLINE**	Medical Literature Analysis and Retrieval System
DE	Definitive endoderm		
DMEM	Dulbecco's modified Eagle's medium	**MEF-CM**	Mouse embryonic fibroblast-conditioned medium
DMSO	Dimethyl sulfoxide	**MEM**	Eagle's minimal essential medium
DSCAM	Down syndrome cell adhesion molecules	**MHC**	Major histocompatibility complex
		MSC	Mesenchymal stem cells
EB	Embryoid body	**MTT**	Methyl-thiazolyl-tetrazolium
ECM	Extracellular matrix		
ESC	Embryonic stem cells	**NIH**	National Institutes of Health
EtOH	Ethyl alcohol		
		PBS	Phosphate-buffered saline
FACS	Fluorescence-activated cell sorting	**PCR**	Polymerase chain reaction
		PECAM	Platelet endothelial cell adhesion molecules
FBS	Fetal bovine serum		
		PPE	Personal protective equipment
GAGS	Glycosaminoglycans	**RPM**	Revolutions per minute
GFP	Green fluorescent protein		

SD	Standard deviation	**TERM**	Tissue engineering and regenerative medicine
SDS	Sodium dodecyl sulfate		
SEM	Standard error of the mean	**TGF-β**	Transforming growth factor beta 1
SMA	Superior mesenteric artery		
STEM	Science, technology, engineering, and mathematics	**TRITC**	Tetramethylrhodamine
		UV	Ultraviolet
TE	Tissue engineering		
TEM	Transmission electron microscope	**VCAM**	Vascular cell adhesion molecules

Reference Search, Image and Data Analysis

Artem Oganesyan and Narine Sarvazyan

Contents

© Springer Nature Switzerland AG 2020
N. Sarvazyan (ed.), *Tissue Engineering*, Learning Materials in Biosciences,
https://doi.org/10.1007/978-3-030-39698-5_1

What You will Learn in This Chapter and Associated Exercises

Students will become familiar with most common databases and Internet search engines that can help them to find and collect information required in this course. They will also learn how to perform basic data analysis and extract quantitative information from acquired images of cells and tissues.

1.1 Literature Search

In today's world where a great amount of data can be digitally stored and readily retrieved, literature search becomes one of the most essential skills that every researcher should own. Below we briefly overview the main search engines that can be used to extract relevant information needed by students to complete their assignments throughout the course.

PubMed is probably one of the best-known free search engines when it comes to biomedical literature. Launched in June of 1997, PubMed gives free access to the Medical Literature Analysis and Retrieval System (MEDLINE)—a large bibliographic database of references, abstracts, and, in many cases, full-text articles on life sciences and various biomedical topics. This database is maintained by the United States National Library of Medicine. As of winter of 2020, PubMed contained over 30 million entries, with half a million new records added each year. One of the biggest assets of PubMed is its search tool. On the left side of the page, users can manage their searches by typing in the category of article, the date of publication, and many other variables. These additional filters, such as language, age of study participants, animal versus human studies, or journal categories, are aimed to narrow down the search. It can be then further improved with sorting articles by "Best Match," "Most Recent," or "First Author" just below the search bar. On the right side of the chosen article page, the user can also see relevant content under "Similar Articles" heading. Once users find the articles that they were looking for, they can send them to a virtual clipboard or store them in "My Bibliography." By using a reference management software (discussed below), the entire content of the user's clipboard can be then added at once to their reference library.

Google Scholar is another widely used free web search engine. Compared with PubMed, it focuses on an even larger array of disciplines, including non-medical sciences (e.g., social sciences, economics, computer science, arts). In addition to this, besides peer-reviewed journal articles, larger literature sources are included in the Google Scholar engine, such as numerous textbooks, conference reports, dissertations, patents, and viewpoints. For beginners, Google Scholar might seem more user-friendly than PubMed since its search tool resembles one of Google's. Nevertheless, searches in Google Scholar tend to be less specific because of the absence of different search strategies and tools provided by PubMed. On the other hand, Google Scholar has a very useful feature (currently absent in PubMed) that shows how many times a particular article was cited and by whom. By clicking on articles that cite the one user just read, one is able to trace later publications on the same subject.

Another rich source of scientific information, including detailed TERM protocols, can be *published patents and patent applications*. They can be found on ▶ USPTO.com website or the more recently developed online platform ▶ www.lens.org. Both provide free access to millions of worldwide patents and are fully searchable. The ▶ Lens.org website also includes access to scholarly articles and enables to use graphics to display results of the searches in a more effective visual way.

1.2 Reference Management Software

Keeping track of multiple publications is a cumbersome task. In order to address this need, several commercial software applications have been developed, common examples being *EndNote* or *Reference Manager*. These reference management programs helped users to collect, store, and manage large numbers of research papers. Yet, these programs were quite expensive, so a group of proactive graduate students from Germany developed a free alternative they called *Mendeley* (after the biologist Gregor Mendel and chemist Dmitri Mendeleyev). Their platform became so popular that in 2013 Mendeley was acquired by Elsevier for over 50 million dollars. One of the sale conditions was that Mendeley continues to be free to users unless a large amount of PDF files is stored with it (to avoid this user can simply unclick "Save PDF" pop-up). One of the most useful functions of Mendeley and other reference management programs is their ability to create a Bibliography (also called Reference list) according to the guidelines of specific journals or publishers. When added to the browser as a Mendeley extension, online papers can be easily imported into a user's reference library to be later inserted in a Word document. Today, Mendeley can run on multiple platforms, including Windows, Mac, iOS, Linux, and Android.

1.3 Literature Access

Most publishing houses do not provide immediate free access to the scientific articles they publish. This is considered by many researchers to be an unfair practice detrimental to the progress of humankind yet very lucrative for scientific publishers. In 2014, the National Institutes of Health (the NIH—an organization that funds most of the biomedical research in the United States for a total annual cost of over 30 billion dollars) required all publishers to release, 12 months after their publication date, the content of the articles that were funded by the NIH. This was a very helpful step, yet it still left a large number of articles unavailable for researchers to read without paying high fees. This list included the most recent ones, the ones that were published before NIH implemented their 12-month policy, and the ones that were not directly funded by the NIH. Overall, out of 14.2 million articles on PubMed that have links to their full-text, only 3.8 million articles are free for any user.

To fight the high cost of article access established by publishing houses, a graduate student from Kazakhstan, Alexandra Elbakyan, started *Sci-Hub*. This website enables free access to scientific articles uploaded there by the users, and as of January 2020, over 78 million papers have been deposited to Sci-Hub [1]. By entering the title, URL, PMID, or DOI of the desired paper in the search bar on the website page, one can download articles in the PDF format bypassing publisher's paywalls. Sci-Hub domains have been periodically suspended due to legal battles with major publishers, and the website address is changing time after time. Yet, by now, Sci-Hub has become a critical working tool for many researchers, especially those in the developing world. In 2016, Alexandra Elbakyan was named by *Nature* magazine to be one of the ten most influential people in science.

Today there is a growing global movement to make science advances available to every person on the planet regardless of their financial status. In August of 2018, the European Research Council decided to make all European scientific publications freely accessible by the year 2021, requiring publishers to remove paywalls and other sorts of obstacles to download journal content (www.coalition-s.org). One can hope that similar trends will be followed by other governmental bodies ushering a new era of scientific information exchange.

1.4 Video Journals

The most convenient way to learn new protocols is to observe them. This can be done remotely by using Youtube, Protocol Exchange, and other searchable video sites. There are also several new journals that publish experimental methods in video format. The most known is called JoVE (*Journal of Visualized Experiments*). JoVE was founded by a frustrated graduate student who tried multiple times to repeat her predecessor's experiments without much luck, and so to address her frustrations she created this new venue. JoVE publishes videos of research protocols from many fields, including tissue engineering. Unfortunately, due to JoVE's overwhelming success, users' access to the videos these days is not always free as JoVE turned into a commercial powerhouse. However, many of the earlier video protocols are still freely available and can be very useful to complete the team's assignments.

1.5 Image Analysis

Statistically valid image analysis is one of the key ingredients to derive proper results and conclusions. Various software programs can help to facilitate this process. One of the most useful and widely used programs was developed by the United States National Institutes of Health (NIH) and is called *ImageJ* [2]. It is free to users and is available on different operating systems, such as Microsoft Windows, MacOS, and Linux. ImageJ has numerous features, and students are encouraged to explore them on their own. There are also multiple online ImageJ tutorials, videos, and articles that explain how a particular parameter can be measured from a digital image. The software allows users to work with different types of images and enables easy editing, analyzing, processing, and storing 8-bit color and grayscale, 16-bit integer, and 32-bit floating-point pictures. ImageJ can read a myriad of various image file formats, including common ones such as PNG, GIF, JPEG, or TIFF or more rare types such as LSM. This can be of particular importance as images obtained from different microscopes and other equipment can have different file extensions.

ImageJ program can help users to measure distances and angles in any given picture or to calculate a specific area within the image. Users can choose certain thresholds for the intensity of selected objects, based on which the program will then collect statistics of different objects' values. After thresholding and object selection, ImageJ can also sort different objects depending on their size, shape, perimeter, or signal intensity (◘ Fig. 1.1). Common geometric affine transformation (e.g., rotation, flips, translation, or scaling) is also possible using ImageJ. The program allows the processing of arithmetical operations between different images. Another useful feature of ImageJ software is its ability to produce histograms, line charts, or 3D rotatable graphs for visual data presentation.

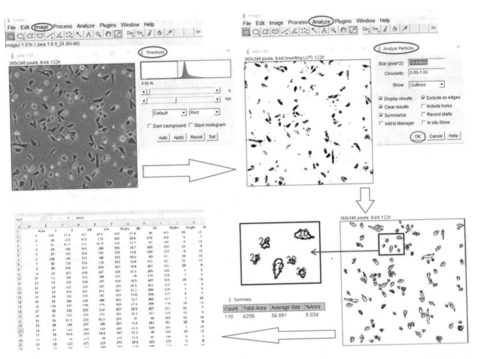

Fig. 1.1 An example of an original and thresholded image of plated cells and a range of different parameters that can be extracted by the ImageJ program from each thresholded object

$$\mu_x = \frac{\sum_{i=1}^{n} x_i}{n}$$

$$SD = \sqrt{\frac{\sum_{i=1}^{n}(x_i - \mu_x)^2}{n-1}}$$

$$SEM = \frac{SD}{\sqrt{n}}$$

Fig. 1.2 Formulas behind basic statistical terms

1.6 Statistical Analysis

Another set of basic skills that students will be required to apply in this course includes a statistical analysis of a given data. Students are referred to various online resources to refresh their memory of basic statistical terms. Here we will just briefly mention the few most commonly used terms (□ Fig. 1.2).

Variance and Standard Deviation (SD) Variance is a principal measure that shows the variability of a quantity. It is calculated by squaring the deviations from the mean and then averaging these squares. Squaring enables one to include not only positive deviations but also negative ones. By taking the square root from the variance, users can calculate the standard deviation (SD), which has the same units as the initial data.

The *Range of a Variable* is the difference between the largest and the smallest given data. Keep in mind that the value of the range depends on the number of experimental points. In other words, the bigger the number of observations, the larger the range of that variable is likely to be.

The *Standard Error of the Mean* (SEM) shows how far the mean derived from N number of samples is from the mean of the entire population. To calculate SEM, first the user calculates the standard deviation as per the formula above and then divides it by the square root of N.

1.7 Understanding the Key Difference Between SD and SEM

It is important for readers to understand the main difference between SD and SEM. Let us say we have a high school consisting of 1200 students. Ten students decided to know their height in absolute numbers. These measured numbers (in centimeters) ended up being: 168; 161; 183; 172; 193; 185; 170; 165; 175; 158. The easiest value to obtain from this sequence is the range of the variable—the difference between the highest and lowest values of our sample, which is 193–158 = 35. In order to know the mean of the sample, we need to sum all these values and divide it by $N = 10$ (called sample size). In our case, the mean will be 173. Once we get the mean, we can calculate SD. The sum of squared deviations from the mean will be 1116. To find SD we will have to divide that number by the sample size minus 1, followed by taking square root from that number. This operation will yield $SD = \sqrt{(1116/(10-1))} = 11.13$. In order to implement identical SD calculations in the Microsoft Excel program, students can use the following formula SD = stdev(data range).

Now, let's say we want to know how far is that sample mean from the mean height of *all* the students in the entire school. The standard error of the mean is then calculated by dividing SD by the square root of $N = 10$ yielding SEM = 3.52. To implement SEM calculation in Excel, one can use the following formula SEM = (stdev(data range))/SQRT(count(data range)).

Looking at the SEM formula (Fig. 1.2), one may conclude that the SEM value will be smaller when more students are included in the calculation. In other words, the more samples that are drawn from the entire school population, the closer the mean will be to the actual mean height value. In our particular example, this value can be ultimately obtained by measuring the height of *all* 1200 students. It will just take more time. In real-life experiments though, measuring *all* the subjects, being them cells, animals, or human patients, is simply not an option. Therefore, by sampling more (i.e., doing more experiments) the user is able to get closer and closer to the ideal mean population value. SEM basically shows how far we are from it.

Note that SEM is directly proportional to the SD or variability of mean values between samples. This means that if the mean heights significantly vary between the students, more sampling (i.e., more N) is required to better estimate a true mean value of a population.

It is important for students to understand the fundamental differences in using these two measures. SD points to how wide are the differences between the individual measurements. SEM shows how close the mean value derived from the multiple measurements is to the mean of the whole population. More about this topic can be found in [3].

1.8 Implications of the Above Statistics for Students' Experimental Design

For future experiments in this course, students should draw conclusions based on at least three independent experiments, each done using at the minimum triplicate measurements. Triplicate means that the same measurement is repeated three times using cells or tissues from the same prep. For example, let's consider how to analyze a possible experiment designed to compare attachment of hepatocytes to laminin versus gelatin-coated dishes. First, a user needs to plate cells on at least six coverslips: three for laminin and three for gelatin group (this will be triplicate measurements for each coating type or $n = 3$). Then the same experiment has to be performed, let's say, four different times using different preps ($N = 4$). Analysis of samples using ImageJ tools enables the user to derive the percentage of cell coverage. The next step is to find average values from triplicates acquired during each of the four experiments (see □ Table 1.1). The next step will be to use the Microsoft Excel program (or its analog) to find mean and SEM for each type of coating. Excel t-test function can then be used to find out whether the two coatings were statistically different. In this case, it will be a paired t-test since each pair of mean measurements (mean values for laminin-coated dishes and mean values for gelatin-coated dishes) was acquired on the same day; therefore, they can be considered "paired." So, the final outcome of this experiment can be worded as: "Attachment of hepatocytes to laminin-coated coverslips was shown to be significantly higher compared to gelatin-coated coverslips. Specifically, expressed as mean±SEM, the percentage of surface cell coverage was 63.92 ± 5.61 vs 33 ± 3.48, $p < 0.05$, $N = 4$ with each individual experiment done in triplicate."

Students are encouraged to use online resources to understand the meaning of other key statistical parameters including correlation coefficient, ANOVA, and different forms of t-test. A good overview of basic statistics for presenting outcomes of the experiments can be found in this freely accessible reference [4].

□ **Table 1.1** An example of an Excel sheet used to calculate the outcomes from four individual experiments ($N = 4$) each done in triplicate ($n = 3$)

Table	Treatment A				Treatment B				T-test
	#1	#2	#3	Avg	#1	#2	#3	Avg	p-value
EXP 1	50	45	24	39.67	67	46	58	57.00	
EXP 2	40	33	34	35.67	44	55	61	53.33	
EXP 3	20	47	33	33.33	81	76	78	78.33	
EXP 4	22	32	16	23.33	45	77	79	67.00	
			MEAN	33.00			MEAN	63.92	0.028
			SD	6.96			SD	11.21	
			SEM	3.48			SEM	5.61	

Session I

Demonstration

The instructor shows students the basic functions of PubMed, Sci-Hub, and Google Scholar as well as an example of how to use Mendeley and its Word plug-in to create a reference list. Then sample images are used to demonstrate key ImageJ functions and how to extract quantitative information and perform statistical analysis using basic Microsoft Excel functions.

Homework

Students are asked to download Mendeley and ImageJ software and familiarize themselves with their main features. Reading assignment: review TERM field history (Kaul & Ventikos, TISSUE ENGINEERING B, v.21, N.2, 2015 or its more recent equivalent).

Session II

Team Exercises

Each team selects a pair of sample images to be quantitatively compared. Any free online publication or other open sources can be used to obtain the two images. These can be cell cultures, decellularized tissues, scaffolds, or any other TERM relevant material. Teams then discuss visual differences between the two images and brainstorm how these differences can be translated into at least three quantitative values using ImageJ analysis tools.

Homework

Teams are tasked with extracting three different quantitative measurements from a selected pair of images using ImageJ tools. A short PowerPoint presentation is made by each team illustrating their quantification process and statistical analysis applied to the extracted data.

🛈 Sample Protocols

Below are examples of ImageJ steps that can be used to analyze images of cells or tissues.

Setting a scale for object measurements (▣ Fig. 1.3)

1. Trace the existing scale bar with the line selection tool.
2. Go to Analyze>Set Scale>Set known distance (e.g., 100) and distance in pixels (line length) and preferable unit (e.g., µm)>ok.
3. Go to Analyze>Tools>Scale Bar>Choose preferable features (width in µm, location, color, font size)>ok.

Counting objects

1. Select an area of interest by area selection tool (▣ Fig. 1.4).
2. Duplicate the image (Image>Duplicate).
3. Minimize the noise by subtracting the background (Process>Subtract the Background).
4. Convert the image into a black-white 8-bit type (Image>Type>8-bit).
5. Set the threshold manually until the cells are well distinguished (Image>Adjust>Threshold).
6. Make the image binary (Process>Binary>Make Binary).

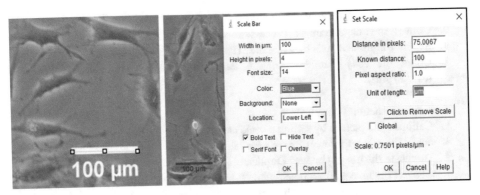

Fig. 1.3 Steps in ImageJ to calibrate an image for extracting quantitative measurements

Fig. 1.4 Use of ImageJ functions to outline and count cells

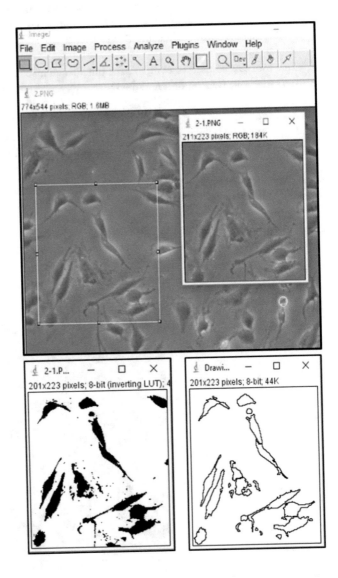

7. Fill holes (Process>Binary>Fill Holes).
8. Convert the image to mask (Process>Binary>Convert to Mask).
9. Separate confluent cells by watershed (Process>Binary>Watershed).
10. Count the cells (Analyze>Analyze Particles>Set a Minimum Particle Size (e.g., 10 μm)).

Area measurement (◼ Fig. 1.5)
1. Calibrate the picture by setting a scale (see above).
2. Select the area of interest by the area selection tool.
3. Duplicate the image (Image>Duplicate).
4. Convert the image into a black-white 8-bit type (Image>Type>8-bit).
5. Adjust the threshold so that the desired area of measurement is black (Image>Adjust>Threshold).
6. Trace the outline of the area of treatment with the brush tool (Click on Wand (Tracing) Tool>Choose brush tool with right-click on oval/elliptical/brush tool).
7. Measure (Analyze>Measure).

This tool can be used to compare the timeline changes of the mean surface area of the same zone (i.e., on different days).

Area fraction
1. Change the type of image to 16-bit (Image>Adjust>16-bit).
2. Set an appropriate threshold (Image>Adjust>Threshold).
3. Set the required measurements: Area, Area fraction, and Display label (Analyze>Set Measurements).
4. Perform the measurement (Analyze>Measure).

The Area fraction tool can be used to compare, for example, the difference in collagen distribution between two sample images (◼ Fig. 1.6).

◼ **Fig. 1.5**   Use of ImageJ functions to manually outline and measure specific areas

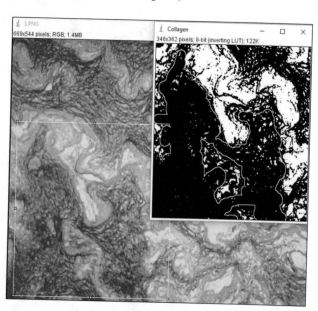

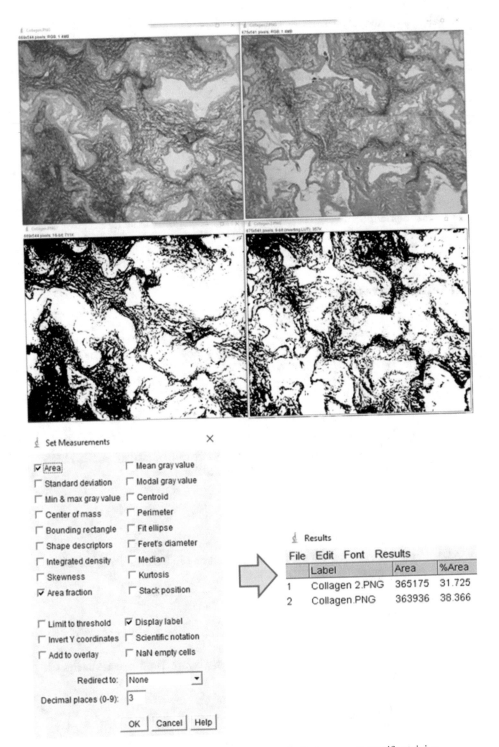

1

Take-Home Message/Lessons Learned ───────────────

After reading this chapter and performing the requested assignments and exercises, students should:

- Become familiar with general tools to search scientific literature including PubMed, Google Scholar, Lens.org, and other online databases
- Learn how to use reference management software, Mendeley being an example
- Become aware of availability of information contained in patents and patent applications
- Familiarize themselves with image processing software such as ImageJ and its basic functions
- Be able to use key statistical formulas to evaluate data and images quantitatively
- Understand the difference between standard deviation and standard error of the mean and its implication for study design

---

Self-Check Questions

? Q.1.1. You've read an interesting article and want to find papers that cited it since they can lead you to more recent work on the same subject. The best free online portal to find them is
  A. Google Scholar
  B. ImageJ
  C. Sci-Hub
  D. JoVE

? Q.1.2. You've read an interesting article and want to find similar papers on the same subject. The best free online portal to find them is
  A. Google Scholar
  B. PubMed
  C. ImageJ
  D. Sci-Hub

? Q.1.3. Free online portal with extensive visualization tools allowing to search for specific protocols within patent texts is
  A. Google Scholar
  B. ImageJ
  C. Sci-Hub
  D. Lens.org

? Q.1.4. Free software with multiple plug-ins enabling comprehensive image analysis is
  A. Google Scholar
  B. ImageJ
  C. Sci-Hub
  D. Lens.org

? Q.1.5. The standard error of the mean is _____ compared to the sample standard deviation.

A. A smaller value
B. A larger value
C. The same value
D. Impossible to tell

References and Further Reading

1. D.S. Himmelstein et al., Sci-Hub provides access to nearly all scholarly literature. elife **7** (2018). https://doi.org/10.7554/eLife.32822
2. C.A. Schneider, W.S. Rasband, K.W. Eliceiri, NIH image to ImageJ: 25 years of image analysis. Nat. Methods **9**(7), 671–675 (2012)
3. P. Barde, M. Barde, What to use to express the variability of data: Standard deviation or standard error of mean? Perspect. Clin. Res. **3**(3), 113 (2012)
4. G. Cumming, F. Fidler, D.L. Vaux, Error bars in experimental biology. J. Cell Biol. **177**(1), 7–11 (2007)

Organ Structure and Vascularization

Mariam Grigoryan and Narine Sarvazyan

Contents

© Springer Nature Switzerland AG 2020
N. Sarvazyan (ed.), *Tissue Engineering*, Learning Materials in Biosciences,
https://doi.org/10.1007/978-3-030-39698-5_2

What You will Learn in This Chapter and Associated Exercises
Students will be briefed about different organ systems, organ ultrastructure, and problems associated with organ transplantation. Students will be required to explore online resources for more details about specific organ of their choice. They will learn differences in organ's vascularization routes and practice cannulating an organ of their choice.

2.1 Main Organs and Their Function

The human body consists of ten organ systems: nervous, cardiovascular, respiratory, digestive, musculoskeletal, endocrine, immune/lymphatic, urinary, reproductive, and integumentary. A failure of any of these systems disturbs the homeostasis and can lead to death. Each organ system usually has one central or key organ. *Heart,* for example, is the central organ of the cardiovascular system, while the *lungs* are the main organs of the respiratory system. *Kidneys* are part of the urinary system and serve to excrete waste from our body with urine. Esophagus, stomach, intestines, liver, and pancreas work with each other as parts of the *digestive system* that takes in food, digests it, absorbs nutrients, and excretes waste with feces. Each organ system is affected by the functions of all other organ systems in the body and is a very complex hierarchical structure, governed by its own homeostatic principles. Organs are made of several types of tissue. Four main tissue types include: *epithelial, connective, muscle, and nervous.* Students who need a refresher or don't have a previous biology background are strongly encouraged to read about the general structure and basic functions of main mammalian organs using available online resources.

2.2 Organ Transplantation

Acute injury or chronic disease can lead to irreversible organ damage. Until recently, the only option for patients with end-stage failing organs was their replacement with healthy organs from another individual. To do so, a healthy organ has to be extracted from a donor and transplanted into a patient, who is then called an organ recipient. Today, the transplantation of most vital organs is possible, including the heart, kidneys, liver, and even lungs [1]. There are three main shortcomings to transplantation. The first is a severe shortage of organs that can be transplanted, so more often than not patients die from organ failure before being matched to a suitable organ donor. Secondly, despite being a life-saving surgery, organ transplantation is not without intraoperative or postoperative complications. Lastly, organ recipients must take immunosuppressive drugs for the rest of their lives, and even that is not guaranteed to prevent the recipient's immune system to reject the implanted organ. Tissue engineering offers a hope to create new healthy organs using a patient's own cells, thus resolving most of the above-mentioned shortcomings [2].

2.3 Tissue Ultrastructure

Each organ contains main cells specific to its function. For example, fundamental functional blocks of the liver are hepatocytes, for the heart – cardiac myocytes, for skin – keratinocytes and fibroblasts, for cartilage – chondrocytes, and so on. Yet, the

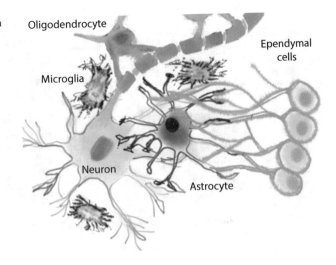

☐ **Fig.2.1.** A cartoon of brain tissue with multiple cell types

same way a skyscraper cannot be built solely from one type of bricks, complex organs cannot be made by simply putting together a large number of cells specific to that particular organ function. In fact, each organ has its own very complex 3D architecture, composed of smaller, yet very complex, functional units, being it nephrons, alveoli, hepatic lobules, etc. ☐ Figure 2.1, for example, shows a cartoon of brain tissue in which one can see at least five types of different cells organized in a very specific manner. To build such functional macroscopic structures is one of the major challenges of the tissue engineering and regenerative medicine (TERM) field.

An analogy one can use to characterize the current state of the TERM field is to compare it to architecture. Today we know how to build a simple cabin from one or two types of bricks. We also know that it should be possible to build a skyscraper since we have seen one (skyscraper being our own body). The challenge is to learn how to use multiple materials to create a much more complicated inner and outer structure of the building with all its parts: plumbing, electrical wiring, disposal routes, heating, etc. working perfectly together. Moreover, such a structure must be adaptable to the environment and be able to self-heal—as our body does. In their efforts to create complex tissues, multiple labs are currently experimenting with layering or printing structures composed of two or three different cell types. Yet, the TERM field is still very far from reaching its ultimate goal of creating functional macroscopic organs made from multiple cell types.

2.4 Vasculature and Its Role

The second major challenge of the TERM field is to provide cells within dense 3D tissue with nutrients while removing cell waste. Evolution was able to solve this task by creating a vast network of tiny blood vessels around almost every cell of the body (☐ Fig. 2.2). The size of the smallest capillaries is so small that ~200 of them can fit into one human hair. Considering this, it is extremely challenging to create a similar network of such fine vessels in vitro. Today a number of promising vascularization strategies is being explored, including seeding scaffolds with endothelial cells

2

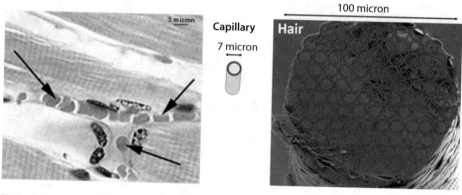

◘ Fig. 2.2 Left: A histology of a muscle showing red blood cells squeezing through thin capillaries between muscle cells. Right: a relative diameter of a capillary versus human hair. Red circles show how many individual capillaries can "fit" inside a single hair

progenitors, 3D printing of small vessels, use of microfluidic chambers, in vivo implantation of multiple thin layers, and many other emerging techniques (more in ► Chap. 11). In addition, each of the red blood cells that squeeze through these capillaries is filled with a highly concentrated solution of hemoglobin that serves as an oxygen and carbon dioxide carrier. The presence of hemoglobin increases the amount of oxygen available to surrounding cells nearly 50 times when compared to a plasma-based solution. So even if vasculature routes are created, continuous perfusion with fluids carrying a sufficient amount of oxygen will be needed to sustain the metabolic needs of engineered tissue that is close to the composition of the native tissue.

2.5 Organ Cannulation

For the purpose of this course, it is important for students to understand the basic anatomy of an animal to be dissected, how to cannulate selected organs to gain access to their major vascular beds, and how to properly use dissection tools. These procedures will become useful for both cell isolation and organ decellularization protocols described in more detail in the following chapters. To isolate live cells, a cannulated organ first undergoes blood removal, followed by collagen digestion using collagenase, trypsin, or other proteolytic enzymes. For organ decellularization, a sequence of detergents in perfusate removes cells, leaving behind only scaffold material. Not every organ can be cannulated, but for organs in which vascular access can be gained, much more efficient cell isolation or organ decellularization can be achieved by proper perfusion.

In general, blood flows from major arteries to smaller arteries, then to arterioles, and finally into what is called a capillary bed. Blood then returns to the heart via a venous system. However, the best way to cannulate a particular organ for perfusion is not necessarily via its main artery. Let's consider three specific cases that illustrate the importance of understanding the vascular anatomy of each organ for its proper cannulation. Specifically, we will discuss cannulation of three different organs of an adult rat, the most commonly available laboratory animal.

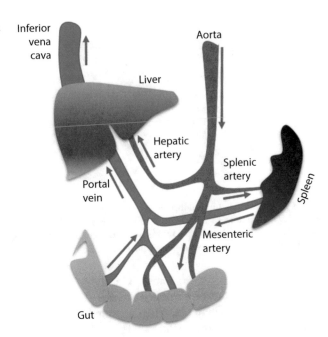

■ **Fig.2.3** A simplified cartoon of gastrointestinal tract circulation showing portal vein being the main vessel to carry blood to the liver

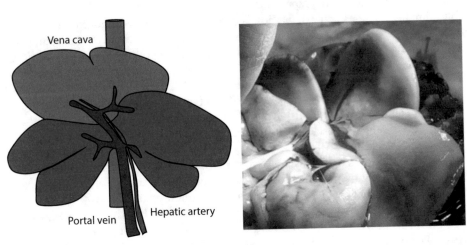

■ **Fig. 2.4** Rat liver cannulation. Left: a cartoon showing the appearance of the portal vein into which cannula has to be inserted. Right: if cannulated properly, liver lobes immediately begin blanching upon blood removal

LIVER cannulation via the portal vein　Most organs can be cannulated through their artery. However, in the case of the liver, cannulation is commonly done via the portal vein. This is because of the peculiarity of hepatic circulation shown in ■ Fig. 2.3. The portal vein is the biggest visible vessel that enters the liver. After opening the animal's abdomen, liver lobes can be flipped upward toward the heart exposing a dark colored portal vein (■ Fig. 2.4). The cannula is then inserted into a small incision within the vein and tied 1–2 mm above the edge of its tip. If cannulated properly, immediately after saline solution starts to flow through the cannula, liver lobes begin to blanch (i.e., lose their dark color).

2

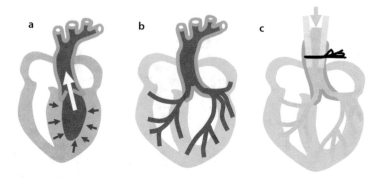

Fig. 2.5 Basics of Langendorff technique to perfuse heart muscle. **a** A cartoon showing how blood is ejected from the left ventricle into the aorta during the systole. **b** During diastole aortic valve is closed, while the muscle is relaxed. This makes aortic blood flow into coronary vessels. **c** When the aorta is cannulated above the aortic valve, perfusate flows into coronary arteries enabling perfusion of the heart muscle

HEART cannulation via the aorta   Coronary arteries supply the heart muscle with blood. The openings of coronary arteries are located right behind leaflets of the one-way aortic valve. Therefore, when a cannula is placed slightly above the aortic valve, fluid pressure keeps the aortic valve closed while pushing flow into coronary arteries. This is called Langendorff preparation [3] (■ Fig. 2.5). In the case of the rat surgery, the animal is first anesthetized, its chest is open, and the heart is removed (■ Fig. 2.6). This is best done by positioning scissors as close as possible behind the heart and cutting it out together with lungs and large vessels. These extra tissues can be then trimmed outside the animal in a Petri dish filled with cold saline. Upon removing these extra tissues, the white arch of the aorta becomes visible due to its size and color. The aorta is then cut down to about 3–4 mm. A cannula is inserted and tied twice using surgical thread. If cannulated properly, immediately after saline solution starts to flow through the cannula, the heart muscle begins to blanch (i.e., lose its pink color).

LUNG cannulation   Lung consists of two systems comprising vascular tissue and airway or respiratory tissue. Therefore, for most efficient access to all lung compartments, both pulmonary artery and trachea are commonly cannulated [4]. This enables inflating the lungs that increase the accessibility of vascular beds to digestive enzymes. For protocols in which cellular content of lungs needs to be simply removed (a process called decellularization to be covered in ▶ Chap. 9), the detergents can be introduced directly through the tracheal route without the need for vascular access. Therefore, for ventilation-based perfusion, the trachea is commonly used and not pulmonary blood vessels.

2.6 Sources of Animal Organs

Different protocols are involved in obtaining freshly excised organs that can serve as sources of live cells. In the case of large agricultural animals, such as cows, sheep, chicken, or goat, a local abattoir is the most common starting place. The excised organ needs to be immediately washed from blood and placed in ice-cold phosphate-

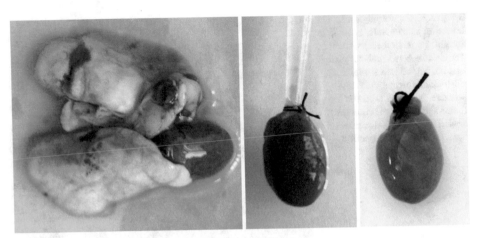

◻ Fig. 2.6 The heart is removed by cutting it out together with lung tissue. After aortic cannulation, lungs and other tissues nearby are trimmed. The heart muscle becomes more pink upon blood washout and continues to contract

buffered saline (PBS) or any other kind of organ-preservation solution. Time to deliver animal tissue to the lab for further steps is crucial for successful cell isolation—the quicker the better. Age, sex, and weight of the animal, as well as tissue collection date and the location of the abattoir, all have to be properly recorded.

In the case of most common small laboratory animals, such as mice, rats, or guinea pigs, tissue or organs are taken from fully anesthetized animals. Necessary steps to minimize stress and pain to the animal must be taken with all the key steps approved by the Institutional Animal Care and Use Committee or its equivalent. Before an organ is extracted, the animal needs to be heparinized (to prevent blood clotting) and anesthetized (to minimize pain and distress). Heparin disrupts blood coagulation cascade preventing the formation of small blood clots that can block sections of vascular beds. Heparin is injected either intravenously or intramuscularly with exact amounts depending on the weight and age of the animal. After heparin injection, the animal is left in an animal cage/container for an additional 10–20 min for heparin to be evenly distributed within the circulation. Then a specified amount of anesthetic agents such as ketamine or pentobarbital is injected. Alternatively, in case of vapor anesthetics, a cotton ball saturated with isoflurane or chloroform can be placed next to the animal while making sure that the container is hermetically closed. In the case of live fish, several drops of clove oil into a water container containing the fish can be used for anesthesia or a standard fish anesthetic such as MS222 can be used. For more detailed information on animal anesthesia, students are referred to open sources such as ▶ https://animal.research.uiowa.edu/iacuc-guide-lines-anesthesia or online reviews [5].

For the purposes of this introductory course, it is expected that a trained instructor or teaching assistant will be conducting any of the above-mentioned procedures involving live animals. Students can proceed with dissection and organ removal only after the animal is fully anesthetized and unresponsive to tail pinch or similar sensitivity tests. Clean and sharp dissecting instruments such as ones shown in ◻ Fig. 2.7 are prepared for dissection including scalpels, scissors, and

Fig. 2.7 Commonly used dissection instruments. From left to right: rat teeth forceps, small straight tweezers, small surgical scissors, curved and straight self-locking forceps, curved tweezers. A scalpel is on the top. Teams should keep their color labeled instruments sharp and clean for each new dissection

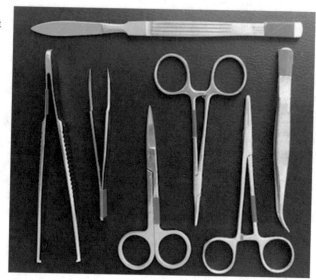

forceps. Surgical scissors are used for general cutting of skin or fascia. They are also useful for spreading tissue layers. Tissue forceps, called "rat teeth," are used to hold thick skin. Instruments must be sterile and so are the gloves of the person who performs the surgery. The surface area can be disinfected using 70% ethanol spray. For dissection, one must know ahead of time the external and internal anatomy of an animal to be dissected. A convenient way to make a custom size cannula for organ perfusion is to cut both sides of the yellow tip at denoted places (**Fig. 2.8**). A narrow side of the cannula is then briefly placed near a candle flame to create a lip. Such a lip enables tying a thread around it preventing its slippage. Having a small custom-fitted wooden or plastic block or platform that holds the syringe with cannula and thread around it can greatly help to perform speedy organ cannulation.

For fish dissection, a line from anus to all the way up to the gills is cut, enabling to expose the internal organs of fish. Note: Due to differences between the anatomy of fish and mammalian species, students should not assume that methods of cannulation described above will work for the fish heart and need to do their homework to understand how to gain access to its vasculature and/or major organ cavities. For rat or mice dissection, pins are used to spatially fix the extremities of the fully anesthetized animal on dissecting tray, after which the abdominal skin area is sterilized using 70% ethanol. Students need to study the anatomy of the selected animal to be dissected ahead of time using available online resources. This will enable them to better understand the relative positions of the organs and blood vessels that lead to them. This is important since the color, the appearance, and the positions of actual organs in a dissected animal can be quite different from what students see in textbook cartoons (**Figs. 2.9** and 2.10).

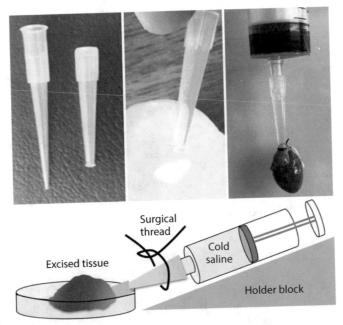

Surgical thread

Cold saline

Excised tissue

Holder block

□ **Fig. 2.8** Cannulating organs using cannulas made from yellow pipette tips. The narrow end of the cannula is cut and flamed to create a small lip. The wide end of the cannula is also cut to enable attachment to a standard-sized syringe. The bottom cartoon shows the positioning of the cannula and thread relative to the vessel to be cannulated. A prism-shaped holder is used to support the syringe in a slightly tilted position

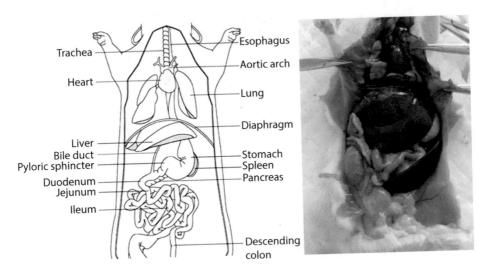

Trachea
Heart
Liver
Bile duct
Pyloric sphincter
Duodenum
Jejunum
Ileum

Esophagus
Aortic arch
Lung
Diaphragm
Stomach
Spleen
Pancreas

Descending colon

□ **Fig.2.9**   A cartoon versus real image of a dissected rat

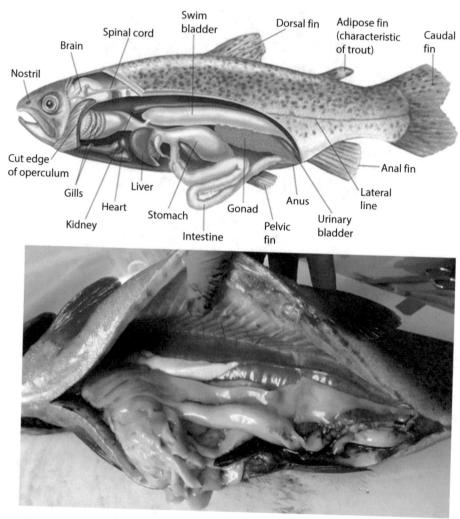

Brain

Nostril

Spinal cord

Swim bladder

Dorsal fin

Adipose fin (characteristic of trout)

Caudal fin

Cut edge of operculum

Gills

Heart

Kidney

Liver

Stomach

Intestine

Gonad

Anus

Pelvic fin

Urinary bladder

Anal fin

Lateral line

Fig.2.10 A cartoon versus real image of a dissected fish

Session I

Demonstration

The instructor shows how to make the cannula, heparinize and anesthetize the animal, followed by cannulation of one of the internal organs. The selection of an animal is based on local availability. For rat anesthesia instructor should follow animal protocol guidelines established by their institution, being it an intraperitoneal injection of ketamine/xylazine, barbiturates, or isoflurane inhalation. To immobilize a live fish, one to two drops of clove oil per liter of water can act as an effective and natural anesthetic. The clove oil should be mixed with a small amount of warm water before adding it to the bucket with cold water containing the fish. Alternatively, fish anesthetic such as MS222 can be used as per the product manual. After a few minutes, the fish becomes motionless and can be dissected.

Homework
Teams are asked to decide on the organ they want to cannulate and read about its function, anatomy, and vascular supply.

Session II
Team Exercises
Surgical area, anesthetics, all required instruments, custom-made cannula, and perfusion solutions have to be prepared by students. After tissue becomes blanched upon blood removal, cannulated organs are placed in the freezer (with cannula attached) to be used later in the course. Alternatively, dissection and cannulation can be done using internal organs of the whole chicken purchased from farmer market or freshly sacrificed large fish. Students are asked to take images throughout the process to document organ appearance before and after cannulation, and after perfusion with saline.

Note - in case of live rat, the animal should be heparinized and anesthetized by the instructor. The following steps can be performed by the students. After the rat is anesthetized, it is important to perform all the surgical procedures rapidly.

Homework
Teams are tasked with the creation of a detailed protocol to describe performed procedures that include images taken at every step of their cannulation experiments.

ⓘ Sample Protocols
Rat liver cannulation
- Check animal responsiveness using tail pinch.
- Fix animal extremities to dissection board using pins or strong tape.
- Sterilize the abdomen with 70% ethanol spray.
- Cut the skin and fascia with surgical scissors.
- Remove any visible hairs not to appear in the working area.
- Stretch and fix the skin and fascia with forceps.
- Open the abdomen to expose the brown-colored liver.
- Gently cut all the ligaments that fasten the organ to other tissues.
- Lift liver lobes up to reveal portal vein, a big bluish-red vessel that goes to the liver's central area.
- Depending on animal weight, the portal vein can vary in size. Several cannulas have to be prepared ahead of time so students can choose the one that fits the best.
- Insert the cannula in vein. Successful insertion is confirmed by seeing blood flowing into the cannula.
- Tie the cannula with surgical thread.
- Carefully lift the liver and cannula out and put them in Petri dish with ice-cold saline for further experimentation.

Ventilation-based rat lung cannulation
- Check animal responsiveness using tail pinch.
- Fix animal extremities to dissection board using pins or strong tape.
- Remove the hair on the chest and abdominal area of the animal.
- Make an incision from the epigastric area through the sternum extending to jugular fossa.
- Open the chest of the animal and locate the lungs, the main bronchi, and the trachea.
- Dissect and remove the lungs together with the main bronchi and the trachea. Be very careful in order to avoid damage and keep these organs intact.
- Identify the trachea. It appears as a small tube with a narrow lumen and cartilaginous consistency. Use a custom-made cannula to attach it to the trachea using a double thread.
- Using a syringe, test if the placement of the cannula is correct and organ is intact. If so, lungs should inflate upon pushing syringe in, and they should shrink back when creating a vacuum by the syringe. Be careful not to damage the lungs with high positive pressure.

Take-Home Message/Lessons Learned

After reading this chapter and performing the requested assignments and exercises, students should:
- Understand the complexity of each organ and main differences between organ systems
- Be familiar with the ways different organs can be cannulated for perfusion
- Be able to make their own cannula and solutions to perfuse organ of their choice
- Know the importance of following approved protocols to minimize stress and pain to the animal used in any experimental procedure

---

Self-Check Questions

Q.2.1. To effectively cannulate and perfuse an organ, one needs to use major _____ leading to that organ
- A. Artery
- B. Vein
- C. Both at the same time
- D. Depends on the organ

Q.2.2. Average capillary has a diameter of about _____
- A. 1 mm
- B. 100 micron
- C. 10 micron
- D. 1 micron

? Q.2.3. To dissociate live cells from the extracellular matrix for future culturing, organs are commonly perfused with
  A. Collagenase
  B. Heparin
  C. Detergents
  D. High calcium concentration solution
  E. Ketamine/xylazine

? Q.2.4. One can anesthetize a rat or mice before dissection using
  A. Collagenase
  B. Heparin
  C. Detergents
  D. High calcium solution
  E. Ketamine/xylazine

? Q.2.5. Blood coagulation can be slowed down by injection of
  A. Collagenase
  B. Heparin
  C. Detergents
  D. High calcium solution
  E. Ketamine/xylazine

References and Further Reading

1. F. Saidi, S.K. Hejazii Kenari, Challenges of organ shortage for transplantation: solutions and opportunities. Int J Organ Transplant Med. **5**(3), 87–96 (2014)
2. T. Welman, S. Michel, N. Segaren, K. Shanmugarajah, Bioengineering for Organ Transplantation: Progress and Challenges. Bioengineered. **6**(5), 257–261 (2015)
3. N. Motayagheni, Modified Langendorff technique for mouse heart cannulation: Improved heart quality and decreased risk of ischemia. MethodsX. (4), 508–512 (2017)
4. T. Tsuchiya, J. Mendez, E.A. Calle et al. Ventilation-Based Decellularization System of the Lung. Biores Open Access. **5**(1), 118–126 (2016)
5. L. Cicero, S. Fazzotta, V.D. Palumbo, G. Cassata, A.I. Lo Monte, Anesthesia protocols in laboratory animals used for scientific purposes. Acta Biomed. **89**(3), 337–342 (2018)

Extracellular Matrix and Adhesion Molecules

Liana Hayrapetyan and Narine Sarvazyan

Contents

© Springer Nature Switzerland AG 2020
N. Sarvazyan (ed.), *Tissue Engineering*, Learning Materials in Biosciences,
https://doi.org/10.1007/978-3-030-39698-5_3

What You will Learn in This Chapter and Associated Exercises
Students will gain a basic knowledge of extracellular matrix components and main classes of cell adhesion molecules. They will then learn how to treat coverslips with different cell adhesion proteins and the key steps of collagen isolation protocol.

3

3.1 The Content and Role of ECM

All tissues consist of cells and extracellular matrix (ECM) that surrounds them. ECM is a complex meshwork of *fibers* (collagen, elastin) and *ground substance*. Ground substance fills the space between cells and fibers and has high water content. During fixation and dying, water evaporates, making ground substance largely invisible on histology slides. The individual components of the ground substance vary depending on the tissue. They include *proteoglycans* (aggrecan, syndecan), *glycosaminoglycans* (dermatan sulfate, heparan sulfate, keratan sulfate, hyaluronan), and *multiadhesive glycoproteins* (fibronectin, laminin).

ECM content depends on the type of tissue. The role of ECM is to provide mechanical and structural support for the cells, give the tissue its appropriate tensile strength, and enable cell communication. ECM components anchor cells through cell-to-ECM attachment adhesion molecules discussed below. ECM also binds different growth factors, such as TGF-b, which are critical for cell growth.

3.2 Cell Adhesion Proteins

Adhesion is a property of cells to attach to surfaces, the latter being other cells, components of the extracellular matrix, natural and artificial scaffolds, or any other surroundings. The ability of the cell to adhere is critical for its differentiation, growth, migration, and even survival. By using different attachment pathways, cells are able to communicate with each other and perform their tissue-specific functions [1].

Points of cell attachment are called cell junctions (■ Fig. 3.1). They have various structures and are responsible for different functions. *Occluding junctions,* also known as *tight* junctions or *zonulae occludentes,* seal off fluid passage between the adjacent cells. *Anchoring junctions* provide mechanical stability and are involved in cell-to-cell attachment, recognition, morphogenesis, and differentiation. *Communicating junctions* allow small molecules to diffuse in and out of the cells as well as between connected cells. They are critical to regulating cell homeostasis.

For the purposes of this course, we will mainly concentrate on anchoring junctions. These types of junctions play a key role in cell-to-cell and cell-to-extracellular matrix adhesion and are made of cell adhesion molecules or CAMs. CAMs are transmembrane proteins containing *intracellular, transmembrane,* and *extracellular* domains. In addition to their role in adhesion, CAMs affect cell migration and proliferation by binding to the intercellular components of the cytoskeleton, which then can trigger multiple downstream pathways.

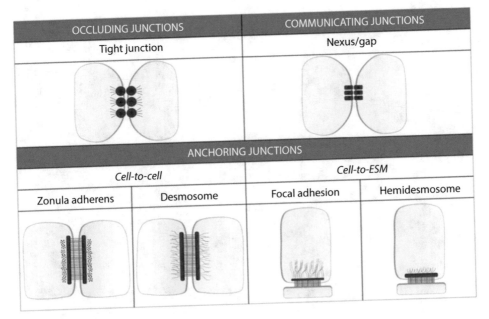

OCCLUDING JUNCTIONS	COMMUNICATING JUNCTIONS
Tight junction	Nexus/gap

ANCHORING JUNCTIONS			
Cell-to-cell		Cell-to-ESM	
Zonula adherens	Desmosome	Focal adhesion	Hemidesmosome

Fig. 3.1 Different types of cell junctions

3.3 Main Classes of CAMs

CAMs are usually classified into four big families: *cadherins, immunoglobulin super-family (IgSF), selectins,* and *integrins* (□ Fig. 3.2). Most of them are Ca-dependent (cadherins, selectins, and integrins). When binding occurs between the same cell adhesion molecules, it is called *homophilic* (cadherins and IgSF); binding between non-identical proteins is called *heterophilic* (selectins, integrins).

Cadherins are Ca-dependent homophilic molecules, which bind to the actin filaments through catenins. There are three subgroups of molecules: N-cadherin (neural), P-cadherin (placental), and E-cadherin (epithelial). The failure of cadherin-cadherin interaction can bring about the development of cancer. This occurs due to the disruption of a pathway called contact-inhibition, which limits the growth of cells connected to their neighbors [2].

Immunoglobulin superfamily (IgSF) plays an important role in inflammation and immune response. This family of adhesion proteins has various subgroups involved in many processes (ICAM – intercellular, C-CAM – cell, VCAM – vascular, DSCAM – Down syndrome, PECAM – platelet endothelial, and many other cell adhesion molecules).

Selectins are mostly found in white blood cells (L-selectin), endothelial cells (E-selectin), and platelets (P-selectin).

Integrins are transmembrane receptors that facilitate cell-extracellular matrix adhesion. They consist of two alpha- and beta-glycoprotein chains, which attach the cells to the components of the extracellular matrix (collagen, laminin, and fibronectin) and actin or intermediate filaments.

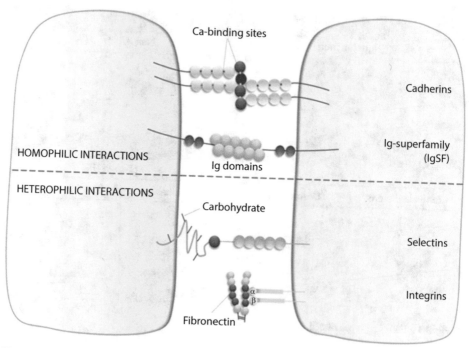

Fig. 3.2 A cartoon showing the main types of CAMs and their interactions

3.4 Attachment Glycoproteins

Multiadhesive glycoproteins stabilize the extracellular matrix, link it to cell surfaces, and connect collagen, proteoglycans, and glycosaminoglycans (□ Fig. 3.3). Here are some of the most important molecules of this class (□ Fig. 3.4).

Fibronectin has several domains; one of them binds to ECM molecules, another one to cell surface receptors. Fibronectin matrix assembly begins when soluble, compact fibronectin dimers are secreted from cells, most often fibroblasts. Fibronectin also participates in the formation of blood clots and is a key protein in wound healing.

Laminin family of glycoproteins is an integral part of the structural scaffolding in almost every tissue of an organism, forming the basis of basal and external laminae. Laminins form independent networks and are associated with type IV collagen networks via entactin, fibronectin, and perlecan. They also bind to cell membranes through integrin receptors and other plasma membrane molecules.

Tenascin is mostly present during the embryogenesis. After this period, it switches off and reactivates only during the regenerative process such as wound healing.

Osteopontin also represents a family of multiadhesive glycoproteins and is mainly present in bones. It binds osteoclasts to the bone surface and plays an important role in calcification of ECM.

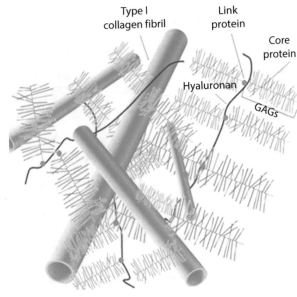

■ **Fig. 3.4** A cartoon showing a simplified structure of an extracellular matrix

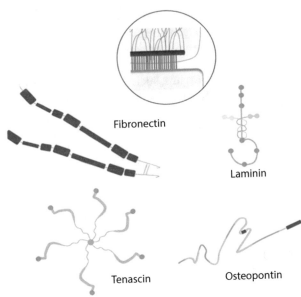

3.5 **Macromolecular Protein Fibers**

Collagen fibers are the most prevalent structure of the extracellular compartment (■ Fig. 3.3). They consist of collagen fibrils, which have different length and width. The structural unit is a collagen molecule, which is a triple helix of three alpha polypeptide chains. If these chains are the same, the molecule is called homotrimeric; if these chains are not the same, the molecule is called heterotrimeric. Currently, twenty-

Fig. 3.5 Histology of epicardial surface of human left atria. Verhoeff-van Gieson staining: collagen fibers show in pink, elastin in black, and muscle in purple. Scale bar: 100 micron

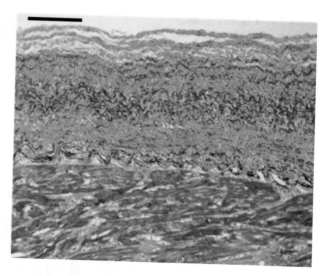

five various types of collagen, which differ in their location and function, have been described. The most abundant type is collagen type I. It is found in skin, bone, tendon, ligaments, dentin, sclera, fascia, and organ capsules and accounts for nearly 90% of body collagen. Collagen fibers can be histologically visible and appear blue when using Masson's trichrome stain or pink when H&E or Verhoeff-van Gieson staining is used.

Reticular fibers also consist of collagen fibrils, but only type III. They can be found in reticular organs, such as the liver, bone marrow, and lymphatic system. The more reticular fibers are found in the tissue, the more mature the tissue is. Reticular fibers can be stained with PAS (periodic acid-Schiff).

Elastin fibers provide tissues with the ability to stretch and return to their original shape when pressure is removed. It is the main component of ECM in the vertebral ligaments, epiglottis, external ear, and elastic arteries. Elastin fibers can be visualized on histology slices using dyes like orcein or resorcin-fuchsin. Staining with Verhoeff-van Gieson shows elastin in black color being coiled into multiple spiral strands (☐ Fig. 3.5).

3.6 Use of Adhesion Molecules in Tissue Engineering and Cell Culturing

After cells are isolated from tissue, in order to survive and proliferate, they need to be attached to flat surfaces of culture plates or to scaffolds in cases of 3D cultures [3]. Since cells do not attach well to plastic or glass surfaces, they have to be coated with adhesion molecules or other substances, which help cells to stick to the surface material. The well-chosen coating agent can greatly affect in vitro cell viability and behavior.

Today, the main sources of coating reagents are different components of native ECM. The most prevalent one is collagen type I. It can be easily isolated from a rat tail or bovine tendons. Hydrolysis of collagen results in the breakup of protein fibrils into a mixture of smaller peptides, called gelatin. Gelatin is probably the most cost-efficient, yet very effective coating agent. Other purified adhesion molecules include laminin and fibronectin. A relatively new approach is to coat cell culture dishes with a crude mixture of ECM proteins derived from decellularized scaffolds [4]. More information about decellularized scaffolds can be found in ▶ Chap. 9.

Negatively charged synthetic peptides, such as polylysine, can also serve as efficient coating agents. They have an additional benefit of reduced contamination risk due to the absence of biological components. Owing to the rapid development of the tissue engineering field, a large variety of commercially coated cell culture plates is now available.

Session I

Demonstration

Students are shown key steps on how to isolate rat tail collagen and to cover plates with solutions of different adhesion molecules such as collagen, laminin, polylysine, or gelatin. The latter are dissolved in PBS at 10–100 µg/mL concentrations, distributed evenly at 50–100 µl/well or glass coverslips, placed in an incubator for 30 minutes, followed by removal of excess fluid and drying under the hood. UV sterilization for 10 minutes can be then used to sterilize coated surfaces. Labeled, sterile coverslips are put into Petri dishes, sealed with Parafilm, and stored in the fridge to be used during subsequent weeks.

Homework

Teams are tasked with reading any recent review article on cell adhesion molecules and their role in creating engineered tissues.

Session II

Team Exercises

Students execute the first part of the collagen isolation protocol. In addition, each team covers cell culture and multiple glass coverslips with solutions of gelatin, collagen, fibronectin, or laminin using steps shown during DEMO session and labels and stores them for the next week's experiment.

Homework

Team members complete collagen isolation protocol and document all the steps involved in treating and storing coverslips. The latter will be used by the teams in the following sessions.

ⓘ Sample Protocols

Collagen preparation from rat tail. The skin of the rat tail is removed with a clamp, and collagen fibers that look like silky white filaments are exposed. Fibers are immersed in ethanol for 3 minutes, wiped dry, and put in a UV-box for 10 minutes. Fibers are then chopped into small pieces and weighed. 100 mL of a 0.1% acetic acid solution is added to per gram of fibers followed by stirring at low speed for 2–3 days at 4 °C. More sophisticated protocols can be found in Reference [5].

Collagen preparation from bovine or porcine Achilles or other major tendons. Tendons are cleaned from any other connective tissues, rinsed with PBS, and

Fig. 3.6 Main steps to cover glass coverslips with adhesion proteins

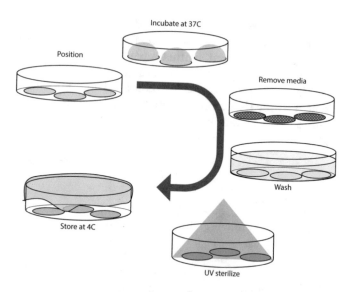

manually chopped or homogenized using a blender with beaker placed in the ice basket to prevent overheating. The pieces are then resuspended in 0.01 M HCl and stirred overnight for 2–3 days at 4 °C.

Samples are centrifuged for 10 minutes at 250 g to remove undigested fibrous material and filtered using a 100–200 μm filter. Sterilized collagen stock solutions can be stored in the fridge for several weeks. More details can be found in Reference [6].

Plate coating (Fig. 3.6)

Reagents and supplies:

- 6-well or 12-well cell culture plates
- #1 Coverslips and forceps
- Incubator or UV-box
- 1% Gelatin solution. Weigh 1 g gelatin and put it into sterile glassware. Add 100 mL distilled water and warm up inside the glassware to get a homogeneous solution.
- If commercial solutions of laminin or fibronectin are available, dilute them in distilled water or as per product manual.

Procedure

- Take coverslips and put them into the dry Petri dish using forceps. Be careful and avoid overlapping of coverslips during the whole procedure. Coverslips can also be put into wells individually.
- Add 10 μg/mL solutions of gelatin, laminin, collagen, or fibronectin on top of each coverslip. Alternatively, fill individual wells.
- Incubate samples for 10–30 minutes at 37 °C.
- Aspirate the solutions making sure not to scratch the coverslips and avoid their overlapping.
- Use PBS or saline to rinse several times.
- Remove the excess of fluid and let dry under the hood.
- Sterilize the covered surfaces under a UV lamp for 10 minutes.

— Take-Home Message/Lessons Learned

After reading this chapter and performing the requested assignments and exercises, students should be able to:
- Name the key components of extracellular matrix
- Distinguish between occluding, anchoring, and communicating junctions
- Identify the main classes of cell adhesion molecules
- List several adhesive glycoproteins and macromolecular protein fibers
- Cover glass coverslips with solutions of adhesive molecules
- Isolate crude collagen fraction

Self-Check Questions

Q.3.1. The main components of the ground substance include the following, EXCEPT
- A. Proteoglycans
- B. Glycosaminoglycans
- C. Cadherins
- D. Glycoproteins

Q.3.2. Choose the correct statement.
- A. Occluding junctions are involved in cell-to-cell recognition, morphogenesis, and differentiation.
- B. Communicating junctions allow small molecules to diffuse in and out of the cells.
- C. Anchoring junctions seal off fluid passage between the adjacent cells.
- D. Selectins are found in all types of cells and tissues.

Q.3.3. The role of ECM is to
- A. Mechanically anchor the cells
- B. Enable communication between different cells
- C. Bind growth factors
- D. All of the above

Q.3.4. These three molecules do not belong to the same class/family:
- A. Cadherin, selectin, integrin
- B. Fibronectin, laminin, osteopontin
- C. Collagen, elastin, immunoglobulin
- D. Dermatan sulfate, heparan sulfate, hyaluronan

Q.3.5. Which protein should not be used for cell attachment to plastic or glass surfaces?
- A. Collagen
- B. Albumin
- C. Polylysine
- D. Laminin

References and Further Reading

1. W.P. Michael, H. Ross, "Connective Tissue," in Histology A TEXT AND ATLAS. With Correlated Cell and Molecular Biology. 156–193 (2015)
2. M. Canel, A. Serrels, M.C. Frame, V.G. Brunton, E-cadherin-integrin crosstalk in cancer invasion and metastasis. J. Cell Sci. **126**(2), 393–401 (2013)
3. A.A. Khalili, M.R. Ahmad, A Review of Cell Adhesion Studies for Biomedical and Biological Applications. Int J. Mol Sci. **16**(8), 18149–18184 (2015)
4. L.T. Saldin, M.C. Cramer, S.S. Velankar, L.J. White, S.F. Badylak, Extracellular matrix hydrogels from decellularized tissues: Structure and function. Acta Biomater. (49), 1–15 (2017)
5. N. Rajan, J. Habermehl, M.F. Coté, C.J. Doillon, D. Mantovani, Preparation of ready-to-use, storable and reconstituted type I collagen from rat tail tendon for tissue engineering applications. Nat. Protoc. **1**(6), 2753–2758 (2007)
6. E. Kheir, T. Stapleton, D. Shaw, Z. Jin, J. Fisher, and E. Ingham, Development and characterization of an acellular porcine cartilage bone matrix for use in tissue engineering. J. Biomed. Mater. Res A. 99 A(2), 283–294 (2011)

Isolating Cells from Tissue

Anna Simonyan and Narine Sarvazyan

Contents

© Springer Nature Switzerland AG 2020
N. Sarvazyan (ed.), *Tissue Engineering*, Learning Materials in Biosciences,
https://doi.org/10.1007/978-3-030-39698-5_4

What You will Learn in This Chapter and Associated Exercises
Students will be introduced to the concept of cell theory and briefed about key intracellular structures. A few examples of cell types that differ in their function and shape will
be then given. Afterward, students will be provided with justification as to why cells need
to be isolated and the main methods to do it. Different ways to count cells after isolation
from tissue will be reviewed followed by hands-on exercises.

4

4.1 Cell Theory

The first person who used the term "cells" was Robert Hooke. In 1665, he used a microscope to examine a thin slice of cork and saw what looked like small boxes. Hooke
called them "cells" because they looked like small rooms that monks lived in. Eight
years later, Anthonie van Leeuwenhoek was first to see living tissue and cells, including
the fertilization process. In 1838–39, Matthias Schleiden concluded that all plants were
made of cells, while Theodore Schwann concluded that all animals were also made of
cells. These individuals are considered to be the cofounders of the "cell theory." Finally,
in 1855, Rudolf Virchow observed, under the microscope, the cell division process.
    Cell theory is based on three postulates: (a) all living things are made of cells; (b)
cells are the basic unit of structure and function in an organism; (c) all cells come
from other pre-existing cells by cell division. Cells make up tissues, tissues are spatially arranged to form organs, several organs working together are considered an
organ system, and all together, this is what comprises a living organism.
    Below we briefly touch upon the terminology of key cell components and differ-
ences in cell shape. Students are referred to multiple free online resources and cell
biology textbooks for more detailed information about specific cellular structures
and their functions.

4.2 Basic Cell Structure

Every cell is different, but there are basic structures that are common to all cells. The
main organelles and their components include plasma membrane, nucleolus, nucleus,
chromosomes, ribosome, vesicle, endoplasmic reticulum, Golgi apparatus, cytoskeleton, cytoplasm, lysosome, centrioles, and mitochondria. The *plasma membrane*, a
microscopic membrane of lipids and proteins, forms the external boundary of the
cell. The *cytoplasm* is a jelly-like substance that fills the cell that contains a multitude
of smaller organelles. *Nuclei* contain genetic material organized as multiple long linear DNA molecules in complex with a large variety of proteins to form chromosomes, which contain all the information that a cell needs to keep itself alive. Inside
nuclei, there are *nucleoli*, which direct synthesis of RNA and form ribosomes.
Endoplasmic reticulum helps move substances within the cell's network of interconnected membranes. There are two types of endoplasmic reticulum: rough endoplasmic reticulum and smooth endoplasmic reticulum. On the rough endoplasmic
reticulum, there are ribosomes attached to the surface; they manufacture proteins
and may modify proteins from ribosomes. The *ribosome* is a complex molecular
machine that serves as the site of biological protein synthesis. Smooth endoplasmic

reticulum does not have ribosomes attached to the surface; there are enzymes that help build molecules, such as carbohydrates and lipids. The *Golgi complex* packages proteins into membrane-bound vesicles. Most of the glycosylation reactions (modification of protein and lipid) occur in the Golgi apparatus. *Lysosomes* are sac-like structures surrounded by a single membrane that holds digestive enzymes. Its function is to break down dying cells, organelles, toxins, and food particles. The *cytoskeleton* is a dynamic, changing structure, made of cytoskeletal filaments: microfilaments, microtubules, and intermediate filaments. The main functions of the cytoskeleton are maintaining and adapting the cell's shape to external influences. The *centriole* is a cylindrical cellular organelle composed mainly from tubulin. The centrioles help the cell to divide (mitosis and meiosis). The *mitochondrion* is a double-membrane-bound organelle, and it has its own DNA. The main function of mitochondria is the production of ATP (adenosine triphosphate), the main molecular unit of cell energy.

4.3 Main Cell Types

All living organisms can be sorted into one of two groups depending on the fundamental structure of their cells. These two groups are the *prokaryotes* and the *eukaryotes*. Prokaryotes are organisms made up of cells that lack a cell nucleus or any membrane-encased organelles. Prokaryotes include bacteria and archaea, two of the three domains of life. Prokaryotic cells were the first form of life on Earth. Eukaryotes are organisms made up of cells that possess a membrane-bound nucleus as well as membrane-bound organelles. Eukaryotic cells are more complex type of cells compared to prokaryotes.

The human body contains over 200 different types of specialized cells. They are derived from three primary germ layers formed during gastrulation, one of the earliest phases of embryonic development. These layers include inner (the endoderm), middle (the mesoderm), and the outer layer (the ectoderm). Eventually, multipotent cells within each layer give rise to all different cells in the body. The inner layer gives rise to cells lining internal body surfaces as well as most of the organs of the gastrointestinal urogenital system. Middle layer cells differentiate to form the heart, the muscle, the bones, the dermis of the skin, and parts of the urogenital system. It also gives rise to bone marrow from which all other blood cell types are derived. Lastly, the ectoderm yields cells of the outermost skin layer, hair, mammary glands, and the cells of both peripheral and central nervous systems.

Each type of differentiated cell is adapted to best perform its function. Below we briefly describe three types of specialized cells to illustrate structural differences that enable their different functions (☐ Fig. 4.1).

Red blood cells, also called *erythrocytes,* are disk-shaped cells of about 7-micron diameter. Their main function is to deliver oxygen to other cells and to remove excess carbon dioxide. In mammals, during maturation process of these cells, erythrocytes lose nucleus giving these cells more room to store oxygen- and carbon dioxide carrier, a protein called *hemoglobin*. This also gives these cells more mechanical flexibility so red blood cells can pass through the smallest capillaries (☐ Fig. 4.2).

Neurons, the main cells of central and peripheral nervous system, have multiple extensions called *dendrites* and *axons*. Places where dendrites and axons connect to

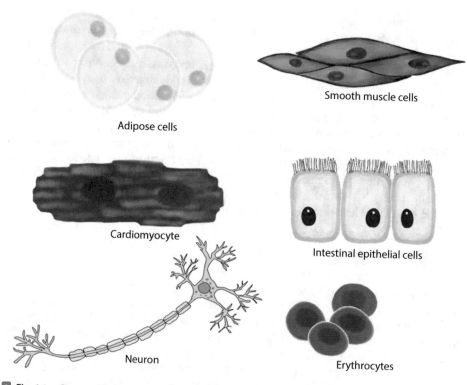

Adipose cells

Smooth muscle cells

Cardiomyocyte

Intestinal epithelial cells

Neuron

Erythrocytes

Fig. 4.1 Cartoon illustrating different shapes of cells with different functions

Fig. 4.2 Transmission electron micrograph of the inner part of cardiac muscle cell showing regular sarcomeres and rows of mitochondria between them. In cardiac myocytes, approximately 30% of cell volume is occupied by ATP producing mitochondria. Scale bar: 1 micron

other neurons are called *synapses*. The role of dendrites is to receive chemical and electrical signals *from* other neurons, while the role of axons is to deliver signals *to* other neurons. Multiple subtypes of neurons exist to perform specialized functions.

Cardiac muscle cells, or *cardiomyocytes,* are big, ~100–150 microns long, brick-shaped cells that are connected to each other through numerous *gap junctions*. The latter enables the rapid spread of excitation through cardiac muscle tissue during heart contraction. The main space within cardiac myocytes is occupied by *sarcomeres*, contractile units of muscle containing actin and myosin. Sarcomeres are surrounded by extensive *sarcoplasmic reticulum* that stores *calcium* ions. Abundant *mitochondria* that provide energy for contraction occupy up to 30% of total cell volume (◻ Fig. 4.2). Several subtypes of cardiomyocytes exist, including ventricular, atrial, Purkinje, atrioventricular, and sinoatrial nodal cells.

These brief descriptions illustrate the major differences in the shapes and intracellular content of matured, fully differentiated cells. Readers are referred to the reference [1] or multiple online resources containing pseudo-colored scanning electron micrographs of different types of cells.

4.4 Why Cells Need to Be Isolated

Tissue engineering implies the process of reconstructing tissue ex vivo from its individual components, that is, extracellular scaffold and cells which make that specific tissue. These individual "building blocks" or isolated cells can be obtained by different means. Some types of cells can be ordered in small amounts from a commercial source and then multiplied by successive culturing. Others can be obtained from progenitor/stem cells, which are first cultured to increase cell amount and then differentiated to the desired cell phenotype. Lastly, many cell types can be isolated from fresh tissues. Crude cell preparations can be then further purified to select a specific cell type needed for a desired tissue engineering step. Alternatively, differential centrifugation, fluorescence-activated cell sorting (FACS), selective pre-plating, or other purification techniques are used to separate cells into different fractions/cell subtypes. These fractions can be then cultured separately and used in different proportions or positioned at different locations when creating engineered tissue.

Isolation of single cells from tissue requires a breakdown of the extracellular matrix that is the structural scaffold supporting cells. The main component of that scaffold is a protein called collagen. Collagenases are a family of enzymes that chop collagen fibers into smaller pieces. Collagenase-based digestion is, therefore, a preferred method to retrieve cells from most of the organs. Specific types of collagenase can be used to receive high yields of cells from certain organs; however, in most cases, crude collagenase works just fine. The second most common proteolytic enzyme, called trypsin, is also often used for non-specific tissue digestion. Other enzymes that can be added include hyaluronidase (to digest glycosaminoglycans) and DNAases (to reduce viscosity resulting from DNA released from damaged cells). In addition to tissue-dissociating enzymes, mechanical methods, such as trituration, can be used. Since calcium ions are an essential part of proteins that connect cells together, the use of calcium-free solutions or chelating agents helps dissociate cells while minimizing damage to their membranes. A balance between overexposing tissues to digestive enzymes (too long, too high concentration) versus having incomplete digestion has to be reached empirically for each type of tissue. Incomplete digestion reduces the number of released cells, called "cell yield." Over-digestion causes damage to cell membranes reducing cell viability.

In order to get digestive enzymes into tissue beds, two main approaches are used. The first one is the use of vascular access. The organ is cannulated and an enzyme-containing solution is allowed to flow through. By flowing through capillary beds, digestive enzymes are distributed evenly within the tissue getting access to nearly every cell. Common examples of cells isolated using perfusion-based approach include cardiac myocytes, hepatocytes, endothelial, and other cell types [2–5]. For tissues or cases when cannulation is not an option, samples are chopped into the smallest possible chunks of ~1 mm^2 and then shaken in enzyme-containing solutions. Notably, manually chopping tissue by sharp scissors was found to be a better approach when compared to the use of homogenizer to break it apart.

4.5 Centrifugation

Suspended cells or cell components can be separated using *centrifugation force* (also known as gravitational force). Tables to calculate required centrifugal speed for each type and size of rotors can be found online. There are also free online calculators such as ▶ https://druckerdiagnostics.com/g-force-calculator.

During the sedimentation process, cell suspension rotates at a high speed and the centrifugal force makes components with higher density or mass to move away from the axis of the centrifuge at a faster speed. The components with lower density or smaller mass move slower and therefore stay closer to the axis. As a result, particles can be separated according to their size or density. Temperature, the density of suspension and particles, rotation speed, and the shape of the particles in the mixture can affect the process. The most common types of centrifugation are differential centrifugation and density gradient centrifugation (◘ Fig. 4.3).

Differential centrifugation is useful in separating out particles with the same or very similar densities, but different sizes or masses. This is because the sedimentation rate is proportional to the square of the particle radius, while only linearly proportional to the difference between its density and density of the media. Differential centrifugation uses a series of cycles with increasing centrifugal speed to produce pellets containing cells or particles with progressively smaller sizes. This is possible because larger components will sediment faster than smaller ones. Therefore, they can be pelleted while the supernatant with smaller particles is collected and centrifuged again at a higher speed. This method is also used to separate different cell organelles and large macromolecules.

Density gradient centrifugation is more complex, but it enables more effective particle separation. Two types of density gradient centrifugation exist. In *rate-zonal centrifugation,* a sample is layered on top of a density gradient. The speed at which particles sediment depends primarily on their size and mass. As the particles in the band move down through the density medium, zones containing particles of similar size form since the faster sedimenting particles move ahead of the slower ones. Because the density of the particles is greater than the density of the gradient, all the particles will eventually form a pellet if centrifuged long enough. In *isopycnic centrifugation,* particles are separated on the basis of their density. The process starts with layering a uniform mixture of a sample on top of density gradient-forming material. When being centrifuged, particles move until their density matches that of the surrounding medium. The gradient is then said to be isopycnic, and the particles are separated according to their buoyancy. For isopycnic centrifugation, the maximum density of the gradient media must be higher than that of the particles, and the

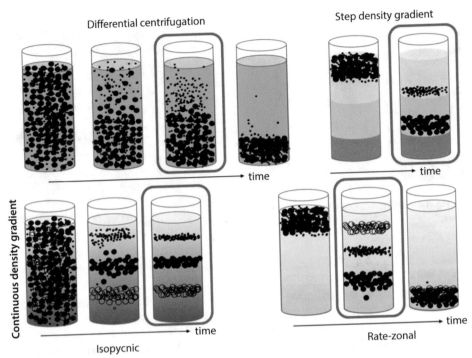

Fig. 4.3 Different types of centrifugation. The red box indicates the time of sample collection. For isopycnic centrifugation, maximal density of the gradient has to be higher than the density of the particles to be separated

end-result is essentially time-independent. *Step density gradient centrifugation* is a modification of the isopycnic separation process. It makes the collection of the sample easier, as particles to be collected are located between the two density gradients.

4.6 Counting Cells

Many methods can be used to count the cells. Here, we will describe the most common ones.

The *manual cell count* method doesn't require much equipment but careful handling. It uses a counting chamber called *hemocytometer* (■ Fig. 4.4). To start, hemocytometer should be cleaned with 70% ethanol and dried with a paper tissue. The next step is to carefully place the special coverslip on the top of the hemocytometer. A small sample of well-mixed cell suspension is then taken using a pipette, and the end of the pipette is gently placed on the edge of the hemocytometer so that cell solution fills the chamber by capillary action. If cell viability is to be determined, a dye called *Trypan blue* is mixed with a cell solution. Trypan blue makes non-viable cells with damaged cell membranes to appear dark blue under microscopic examination. To calculate the total number of cells, the following formula is used: cell count /number of quadrants counted x dilution factor × 10,000 × cell suspension volume (in mL) = total cell yield. Cell viability can be then calculated as a percent of live cells versus total cell numbers. In case

4

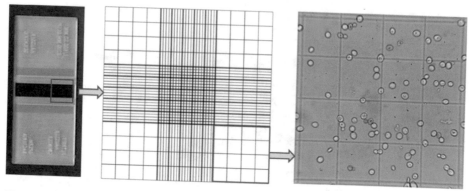

Fig. 4.4 Cell count using hemocytometer. Trypan blue stains dead cells blue while live cells appear clear

shown in ● Fig. 4.4, it will be 68 clear cells/(68 clear cells +10 blue cells) = 68/78 x 100% = 87%.

More advanced cell counting methods are based on the use of *automated cell counters*. These devices perform multiple counts of cells within a known area and average out the results. They are both precise and reliable but may have difficulty obtaining accurate counts when cells are irregularly shaped or in case of very diluted samples.

The *Coulter counter* is an electronic particle counter that provides an alternative to the hemocytometer for counting cells. Cells in suspension pass through an aperture across which flows an electric current. Their presence alters the electrical resistance of the medium and induces changes in the current flow and voltage. The magnitude of these changes is directly proportional to cell size and is electronically converted to a particle count.

Fluorescent cell counters detect and count live and dead cells with the help of specific fluorescent dyes such as acridine orange and propidium iodide. The advantages of fluorescent cell counters include their ability to distinguish non-cellular debris from cells and/or to select a specifically labeled cell population.

The *flow cytometry* approach uses *FACS* (*fluorescence-activated cell sorting*) machines designed for detailed and specific cell analysis. They can differentiate between cells of similar sizes based on the expression of specific surface proteins and provide fast, objective, and quantitative recording of fluorescent signals from individual cells. FACS is an extremely lengthy and expensive method to be used for a simple cell count. Instead, it is widely used as a powerful research tool to separate different cell populations based on their size, shape, and specific protein expression.

Spectrophotometry can be used to estimate different quantities of particles in suspension based on how much light is absorbed by a cell-containing solution. A cuvette with more concentrated suspension will absorb more light indicating higher cell density. Spectrophotometer-based approach does not give absolute cell count, and it doesn't distinguish viable cells from other particles that might also absorb passing light. Yet it can be an effective tool to compare cell density of different samples. For a quantitative estimate, it will require a standard curve based on absorption values from cell suspensions counted by other means.

Session I

Demonstration

The instructor shows how to isolate neonatal rat ventricular myocytes and cardiac fibroblasts (■ Fig. 4.5). Briefly, hearts are excised from 1- to 2-day old rats, rinsed in cold PBS, and minced into ~1 mm³ pieces. Tissue pieces are then incubated with papain and thermolysin enzyme mixture for 30–40 min at 37°C (Pierce kit #88281). Alternatively, solutions of collagenase II or trypsin can be used (0.1% trypsin and 0.05% collagenase II are suggested starting concentrations that will need to be adjusted based on enzyme activity or units). After 30 min incubation with enzyme solution, tissue chunks are gently triturated and passed through a cell strainer to remove any undigested pieces. Cells are then washed several times with Dulbecco's Modified Eagle Medium (DMEM) and counted (■ Fig. 4.5).

The first-order separation of cardiac fibroblasts from myocytes can be done using a so-called preferential attachment protocol (■ Fig. 4.6). Fibroblasts are much better at attaching to the bottom of the flask after 1 h incubation in cell culture incubator. Most of the myocytes, on the other hand, are poorly attached and can be detached by gently swirling the flask. The media with suspended myocytes is then collected and plated on glass coverslips for next week's demo. Fibroblasts can continue to grow within the flask until they reach confluency (more in ▶ Chap. 5).

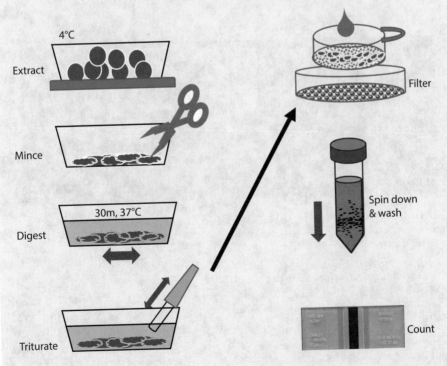

■ **Fig. 4.5** Main steps to isolate cardiac cells from neonatal rats (phase I)

4

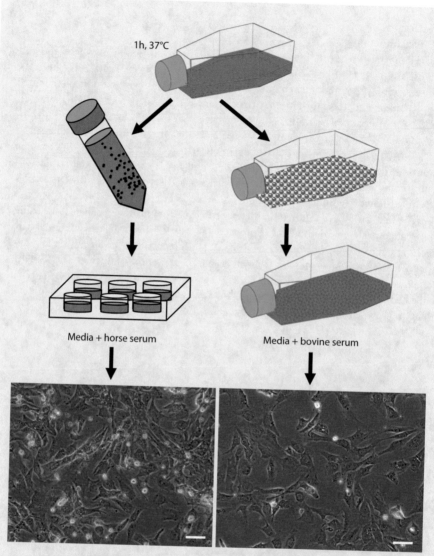

1h, 37°C

Media + horse serum

Media + bovine serum

🔲 **Fig. 4.6** Top: main steps for the first-order separation of myocytes from fibroblasts. Bottom: representative images showing the appearance of myocytes (left) and fibroblasts (right) after 3 days in culture. Scale bar: 50 micron

Preferential attachment is not a perfect way to separate these two cell types. Within a few days after isolation, beating myocytes will be present and can be visually observed in flasks containing fibroblast cultures. However, once fibroblasts are re-plated, the number of myocytes will dramatically decline. Some fibroblasts will also be always present in cardiac myocytes cultures. The use of horse serum instead of bovine serum inhibits fibroblast growth while promoting the survival of myocytes.

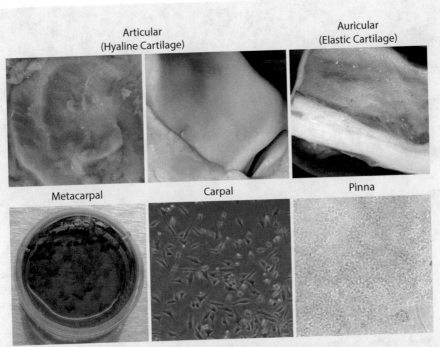

Fig. 4.7 Images from protocol describing the isolation of bovine chondrocytes. Top: cartilage sources. Bottom: chopped pieces of cartilage in collagenase solution, isolated plated chondrocytes, verification of chondrocytes phenotype using Alcian blue staining. The latter stain is a common marker for extracellular matrix proteoglycans produced by chondrocytes

The latter should form confluent layers of visibly contracting cells starting from day 3. This self-beating behavior will continue for about a week after which the quality of myocyte layers will decline. Fibroblast cultures, on the other hand, can be passaged several times and used during subsequent weeks for a demo or student exercises.

In the absence of vivarium, isolation of aortic smooth muscle cells, fibroblasts, or endothelial cells can be used to illustrate cell isolation using trypsin-based digestion of aortic wall layers. Alternatively, isolation of chondrocytes from the cartilage of any major joint can be shown (☐ Fig. 4.7). Freshly excised porcine, bovine, or sheep tissues brought to the lab on ice from a local abattoir can be used for the latter procedures.

Note: If the commercial cell line is not available, it is advised to use the obtained suspensions of fibroblasts, endothelial cells, or chondrocytes to create primary cell cultures. These cell types should reach confluency within a week or two and can be used for demo sessions for the following weeks.

Homework

Teams are tasked with searching the literature to find the simplest and most suitable detailed protocol to isolate cells from their organ of choice.

Session II

Team Exercise

Based on their homework assignment, each team attempts to isolate cells based on either dispersion or perfusion-based protocol followed by collagenase or trypsin digestion, centrifugation, and cell count. Isolated cells are then plated on coverslips prepared by the students during the previous session.

Homework

Teams are tasked with a detailed description of performed cell isolation steps, including phase-contrast images of the cells, and calculation of cell yield and viability.

ℹ️ **Sample Protocols**

Rat hepatocyte isolation. Follow steps from Sample Protocols in ▶ Chap. 2 to cannulate rat liver. Shave the rat's abdomen, disinfect with alcohol, and immobilize it on a surgery table. Open the abdomen. Find a portal vein and insert a cannula. Perfuse with saline solution until the blanching of the liver. Perfuse with 0.3% collagenase II solution. Dissect the liver. Disconnect the perfused rat liver from the perfusion apparatus and transfer it to a Petri dish filled with low calcium media. Cut tissue into small pieces. Filter the resulting cell suspension through a sterile filter. Centrifuge the cell suspension for 10 min at 1500 rpm. Remove the supernatant and wash cells with complete media. Centrifuge cell suspension 10 min at 1500 rpm. Filter the resulting cell suspension through a sterile filter.

Chondrocytes from bovine articular or auricular cartilage. Pieces of bovine or other large farm animal cartilage can be dissected and transferred to the lab in an ice-cold solution. Thin slices can be scraped off using a scalpel, or just cut into 5 × 5 mm pieces. Enzymatic digestion is performed using 0.3% collagenase II with samples stored in cell incubator overnight (37 °C, 5% CO_2). Complete medium (DMEM) is added to stop the action of collagenase. After filtration of digested tissue, samples are centrifuged at 2400 rpm for 10 min. The supernatant is discarded and the pellet is resuspended in 5 mL of complete medium. The last step is repeated twice.

Endothelial cells from pig aorta. Long pieces of freshly cut pig aorta, 1–4 cm in size, should be placed in ice-cold saline and transferred to the lab within 3 h from excision time. The segments should be then pinned to the board with the intima layer up. Using a sharp blade at 45° angle surface of the segment is gently scraped in a longitudinal direction (in relation to the vessel's main direction). After each scratch accumulated tissue is transferred into 50 mL tube pre-filled to 25–30 mL of cell culture media (M199, DMEM, or similar). The content of the tube is then centrifuged at 250 xg for 10 min. The supernatant is discarded, while cell pellet is resuspended in fresh media. The last step should be repeated twice.

Cell count. After cells are isolated, students must check the yield and viability of cells using Trypan blue exclusion test by mixing 10 μL of cell suspension with 10 μL 0.25% Trypan blue in PBS followed by the count of viable (clear) and dead (blue) cells as per ◻ Fig. 4.4.

```
┌─────────────────────────────────────────────────────────────────────────┐
```
Take-Home Message/Lessons Learned

After reading this chapter and performing the requested assignments and exercises, students should:

- Understand the reasons why cells need to be isolated and how to do it with minimal damage to cell integrity
- Know what enzymes are commonly used for cell isolation
- Understand the main differences between various methods to count the number of isolated cells
- Be able to differentiate between step density, isopycnic, and differential centrifugation
- Be able to calculate cell numbers using hemocytometer

Self-Check Questions

? Q.4.1. It is critical to lower the concentration of _____ions when isolating cells from tissues.
A. Sodium
B. Calcium
C. Potassium
D. Magnesium

? Q.4.2. An enzyme that is NOT commonly used for isolation of cells from tissues is
A. Collagenase
B. Trypsin
C. ATPase
D. DNAase

? Q.4.3. The best way to separate cells based on their density is by using
A. Flow cytometry
B. Isopycnic centrifugation
C. Trypan blue staining
D. The Coulter counter

? Q.4.4. Any of the listed approaches can be used to estimate cell sizes, EXCEPT
A. Trypan blue staining
B. Flow cytometry
C. The Coulter counter
D. Differential centrifugation

? Q.4.5. Correct timing to collect sample is critical when performing_____ centrifugation.
A. Step-density gradient
B. Isopycnic
C. Differential

References and Further Reading

1. R. Bailey, Different cell types in the body. [Online]. Available: https://www.thoughtco.com/types-of-cells-in-the-body-373388. Accessed 13 Jun 2018
2. M.-L. Izamis et al., Simple machine perfusion significantly enhances hepatocyte yields of ischemic and fresh rat livers. Cell Med. 4(3), 109–123 (2013)
3. M.N. Berry, D.S. Friend, High-yield preparation of isolated rat liver parenchymal cells: a biochemical and fine structural study. J. Cell Biol. 43(3), 506–520 (1969)
4. T. Powell, V.W. Twist, A rapid technique for the isolation and purification of adult cardiac muscle cells having respiratory control and a tolerance to calcium. Biochem. Biophys. Res. Commun. 72(1), 327–333 (1976)
5. D.G. Leuning, E. Lievers, M.E.J. Reinders, C. van Kooten, M.A. Engelse, T.J. Rabelink, A novel clinical grade isolation method for human kidney perivascular stromal cells. J. Vis. Exp. 126 (2017). https://doi.org/10.3791/55841

4

Functional Assays and Toxicity Screening

Hovhannes Arestakesyan and Narine Sarvazyan

Contents

© Springer Nature Switzerland AG 2020
N. Sarvazyan (ed.), *Tissue Engineering*, Learning Materials in Biosciences,
https://doi.org/10.1007/978-3-030-39698-5_5

What You will Learn in This Chapter and Associated Exercises

Students will be taught how to evaluate viability of isolated cells using several widely accepted methods. Cell type–specific functional assays will be then briefly discussed. Concept of the standard curve, blank, and positive and negative controls will be then introduced, followed by practical exercises to test toxicity of different compounds using these concepts.

5.1 Assessment of Cell Viability

Viability assays are used to determine how many viable cells are present after their isolation from the tissue, and how these cells survive in culture under different culturing conditions. They are also used to assess the success of routine steps such as cell passaging, cryopreservation, or thawing (more in ▶ Chap. 6). During the 3D printing process, evaluation of cell viability is important during multiple time points, that is, *before* cells are mixed with bioink material, *after* extrusion and crosslinking, and *during* subsequent tissue formation. Lastly, viability assays are widely used to evaluate the toxicity of different drugs or treatments. Most viability assays can be modified to work for cells in suspension, cultured cells, as well as for cells residing with scaffolds of engineered tissue.

Based on the specific end goal, viability assays rely on different aspects of cell metabolism, being it mitochondrial activity, protein turnover, ATP production, DNA replication rate, and others [1]. Regardless of the chosen assay, there are many indirect factors that can influence the assay's outcome. These factors include possible effects of media components, cell surface-to-reagent volume ratio, fluid evaporation, and many others. Before choosing an assay, students have to understand exactly what information has to be quantified, being it the number of live/dead cells (viability/cytotoxicity assay), the total number of cells (proliferation assay), or the exact mechanism of the cell death (e.g., necrosis, apoptosis). The simplest method to examine viability is to stain cells with Trypan blue as described in ▶ Chap. 4. Trypan blue is a synthetic dye that selectively stains cells with compromised membranes. Upon entry into the cell, Trypan blue binds to intracellular proteins, thereby rendering cells dark. Cell viability can be then calculated as the percent of live cells versus total cell numbers as shown in ◘ Fig. 4.4.

5.2 The Most Commonly Used Assays to Evaluate Overall Cell Viability: DAPI, Ethidium Bromide

Another method of distinguishing dead or dying cells is the use of DNA binding fluorescent dyes. These dyes are impermeable through the cell membrane when cells are live; therefore, they can be used as fluorescent indicators of dead cells. There are many fluorescent dyes such as propidium iodide (PI), ethidium bromide (EB), 7-aminoactinomycin D (7-AAD), 4′,6-diamidine-2-phenylindole (DAPI), and acridine orange (AO) that can be used on live cells without fixing. Upon interaction with nucleotides, these dyes exhibit intense fluorescence. DAPI molecules attach at the minor groove of the DNA double helix. EB intercalates inside the DNA double helix.

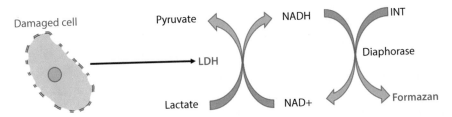

Fig. 5.1 Chemistry behind the LDH cytotoxicity assay

LDH assay Lactate dehydrogenase (LDH) is a cytosolic enzyme found in all living cells. LDH cytotoxicity assay measures the release of LDH into the media, which occurs when the cell plasma membrane is damaged. To quantify the amount of released LDH, the following enzymatic reaction is used. First, LDH catalyzes the conversion of lactate to pyruvate acid, as it converts NAD^+ to NADH. Second, diaphorase uses NADH to reduce a tetrazolium salt to a red formazan product (🔲 Fig. 5.1). Therefore, the level of formazan formation is directly proportional to the amount of released LDH in the medium.

Resazurin reduction assay Resazurin sodium salt is a cell-permeable non-fluorescent dye that can be used to monitor the number of viable cells. After adding blue colored resazurin in the media, viable cells reduce it into the pink and fluorescent resorufin. The produced resorufin quantity is proportional to the number of viable cells and can be quantified either by measuring a red fluorescence using a fluorometer (Excitation/Emission 560/590 nm) or by measuring a change in absorbance at 570 nm (🔲 Fig. 5.2).

MTT Tetrazolium reduction assay The methyl-thiazolyl-tetrazolium assay is a standard colorimetric assay used to assess the metabolic activity of cells. It measures the reduction of yellow 3-(4,5-dimethylthiazol-2-yl)-2,5-diphenyl tetrazolium bromide

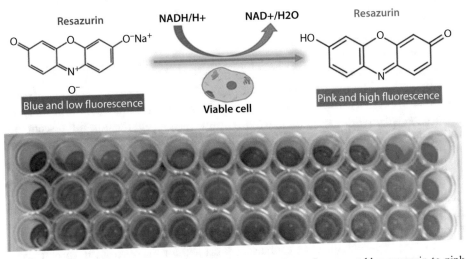

Fig. 5.2 Resazurin assay. Top: intracellular conversion of non-fluorescent blue resazurin to pink fluorescent resorufin. Bottom: an image of a cell plate showing the conversion of blue resazurin to pink resorufin with triplicates for each condition with cells progressively losing their viability from right to left

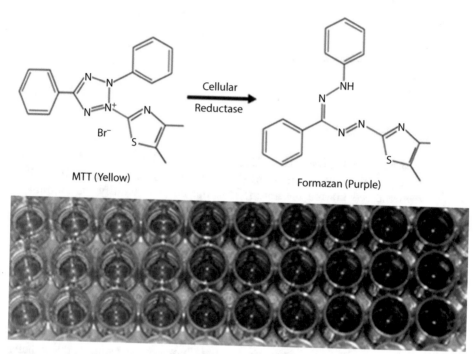

Cellular Reductase

MTT (Yellow)

Formazan (Purple)

◻ Fig. 5.3 MTT assay. Top: formula showing the conversion of MTT to formazan upon the action of intracellular reductases. Bottom: image of a cell plate after MTT assay showing the conversion of a clear solution to purple with three samples for each condition with cells progressively losing their viability from right to left

by mitochondrial succinate dehydrogenase. The MTT enters the cells and passes into the mitochondria where it is reduced to an insoluble, colored formazan (dark purple) product (◻ Fig. 5.3). To dissolve precipitated formazan crystals, MTT solvent should be added. The quantity of formazan is measured by reading absorbance at 570 nm using microplate spectrophotometer.

LIVE/DEAD assay   A number of commercially available viability assay kits use two-color fluorescent reagents to discriminate the population of live cells from the dead cell population. The most commonly used LIVE/DEAD cell assay measures intracellular esterase activity using properties of *calcein-AM* and plasma membrane integrity using *ethidium bromide*. Esterase activity removes esters from the non-fluorescent and cell-permeable form of calcein-AM dye converting it to a highly fluorescent non-ester form. The latter has multiple charges and as such is well retained within live cells. It yields an intense cytosolic fluorescence with excitation/emission maxima of 495 and 515 nm, respectively. Ethidium bromide enters cells with a damaged membrane. Once inside the cell nuclei, it undergoes a ~40-fold enhancement of fluorescence upon binding to nucleic acids, thereby producing a bright red fluorescence in dead cells with excitation/emission maxima of 495 and 635 nm, respectively (◻ Fig. 5.4).

Knowing the mechanism of each assay is critical for the correct interpretation of its results. For example, use of calcium-AM staining to evaluate the viability of cells upon treating them with a compound that affects intracellular esterase activity, while not necessarily killing these cells, can yield an erroneous conclusion. This is because

Fig. 5.4 Culture of mouse fibroblasts stained with LIVE/DEAD Cytotoxicity Assay Kit. Live cells exhibit bright green color due to the presence of calcein in the cytosol, whereas dead cells with compromised membranes have red-stained nuclei

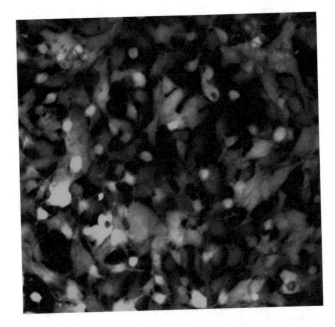

intracellular esterases are involved in removing ester moieties of calcium-AM dye, which leads to trapping dye within the cell. One has to also keep in mind that cell viability is a rather general term that can include cytotoxic, cytostatic, and antiproliferative effects.

5.3 Controls and Linear Range

Another important issue to consider is the design of appropriate controls. Many factors can affect measurement endpoints, so one needs to make sure that outcomes of viability assays are not mistakenly attributed to other factors. There are three types of controls that can be used in cell viability assays. Those include positive control, negative control, and blank. *The positive control* is a sample that for sure will be affected by the treatment. For example, high concentration hydrogen peroxide can serve as a positive control. *The negative control* is a sample w/o treatment. Lastly, *the blank* means sample without any cells but with the media. The latter is needed to subtract the values of background fluorescence or absorbance for all the samples.

It is very important to establish the linearity of the assay curve as per ◘ Fig. 5.5. For each assay it is advised to set up a standard curve, which will allow to later transform data into a concentration-response curve. In the case of LDH assay, it can be a concentrated LDH solution; in case of resazurin, it can be a resorufin standard. The measurement data obtained from the treated samples should be compared to a standard curve of serially diluted standards specific for each assay.

To ensure the reproducibility of the measurements, it is recommended to perform all types of viability assays using at least three samples, that is, *triplicates*. The average of the three can then be used as a measure of viability for each sample in that particular experiment. The experiment itself then needs to be performed at least three times (see ▸ Chap. 1).

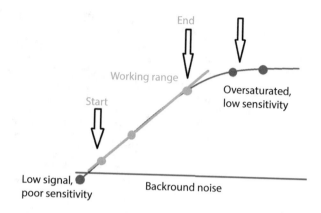

◘ Fig. 5.5 Optimal results are obtained when sample values are found within the linear range shown in green

5

5.4 Functional Assays

The above viability assays are applicable to almost all cell types. The second set of assays is more specific as is used to evaluate a particular function of specific cell types (◘ Fig. 5.6, *top panel*).

For example, the dynamic monitoring of intracellular calcium is a standard way to monitor the functional performance of muscle cells. This is done by loading cells with live calcium indicators, such as Fluo-3, Fluo-4, Fura, and many others. Upon binding of calcium ions to these dyes, their fluorescence dramatically changes. Calcium transient is a momentary >10-fold increase in the concentration of cytosolic calcium (Ca^{2+}). It occurs due to the rapid release and re-uptake of calcium ions from the sarcoplasmic reticulum. This transient increase in Ca^{2+} triggers the contraction of both cardiac and skeletal muscle cells. Quantitative indices extracted from recordings of calcium transients, such as transient amplitude, half times of upstroke and decay, and synchronicity, provide a great deal of information regarding the expression of specific channels in these cells. Voltage-sensitive dyes are used to record the activity of electrically active cells. These cells also called excitable cells and include cardiac, skeletal, and smooth muscle cells together with neurons.

Other types of functional assays are aimed at labeling the expression of particular genes or the production of different proteins coupled to a specific cell phenotype. For example, mature hepatocytes are expected to produce significant amounts of albumin; functionally active chondrocytes should deposit *glycosaminoglycans* (GAGs), fibroblasts, collagen, etc. Production of specific proteins or expression of specific genes can be monitored on both cell and tissue constructs levels. But there are also functional assays that only work for engineered tissues exceeding mm scale (◘ Fig. 5.6, *bottom panel*). In the latter case, assessment relies on properties of engineered tissue that consists of many interconnected cells in close interaction with surrounding ECM or artificial scaffold material. Such macroscopic tissues can be stretched, compressed, bent, and electrically stimulated to produce measurable active force or tested for their barrier function (in case of tissue constructs emulating the skin).

◻ Fig. 5.6 Summary of different types of assays that can be used to evaluate survival and specific functions of cells and tissue constructs

CELLS

Viability assays applicable to most cell types	Functional assays applicable to specific cell types
Trypan Blue LDH, resazurin, MTT, LIVE/DEAD, ATP production, FACS	Expression of certain genes Production of specific proteins Changes in intracellular calcium Electrical activity/action potentials

TISSUE CONSTRUCTS

Viability assays applicable to most tissue types	Functional assays applicable to specific tissue types
Bioluminescence, LIVE/DEAD, resazurin, LDH, TTC, MTT assays Presence of cells (histology)	Contractile strength, stiffness, elasticity, microarchitecture, ECM formation, barrier function, vascularization, mineralization

Session I

Demonstration

The instructor exposes plated cells to different concentrations of hydrogen peroxide followed by either resazurin, LDH, or MTT assay. Recommended doses of H_2O_2 for 10 min exposure include 0.1 mM, 1 mM, 10 mM, and 100 mM peroxide. Assessment of cell contractions and other tissue-specific functional assays for toxicity assessment is also demonstrated and discussed. For example, the use of Fluo-4 to monitor calcium transients using plated rat neonatal cardiomyocytes from the previous week can be shown.

Homework

Based on a literature search, each team designs an experiment in which passaged cells from the previous week can be exposed to chemical or physical stress followed by cell viability assessment.

Session II

Team Exercises

Teams test chosen stress conditions and methods to determine cell viability afterward. Assay manufacturers provide a detailed description of each assay. Students are advised to carefully read these manuals before experiments. Their goal should be to create a viability curve that should have at least three points control (i.e., no damage), mid-point (some damage), and max damage (i.e., all cells dead) points. Cultured cells from the previous week or commercial cell lines can be used to conduct these experiments.

Homework

Teams use their experimental data to create a graph illustrating the effect of their chosen treatment on cell viability. Positive, negative, and blank samples must be included.

ⓘ Sample Protocols

DAPI staining protocol

1. Remove culture media and rinse the sample with PBS.
2. Dilute dye stock solutions with PBS to 300 nM final concentration.
3. Add the appropriate volume of the diluted staining solution to the prepared coverslip. Make sure that the cells are completely covered.
4. Incubate for 5 min.
5. Wash the sample three times with PBS. Drain excess buffer and mount with Mowiol mounting medium.
6. Use a fluorescence microscope with appropriate filters to view the sample.

Protocol for LDH assay

1. Transfer supernatant from cell culture plates or treated cells into a new multiwell plate.
2. Add LDH reaction mixture according to manufacturer suggestions.
3. Incubate at room temperature for 30 min.
4. Add stop solution to stop the enzymatic reaction.
5. Measure absorbance with a spectrophotometer at 490 nm.

Protocol for resazurin assay

1. Add the specified amount of resazurin solution to the samples (in case of 96 wells this can be 10 µL of 0.15 mg/mL resazurin to the well containing 90 µL of cell culture media).
2. Incubate for 2–4 h at 37 °C in the dark.
3. Record fluorescence using a 560/590 nm (excitation/emission) filter set, or absorbance at 570 nm.

Procedure for MTT assay

1. Add a specified amount of MTT solution to the samples (in case of 96 wells this can be 10 µL of 5 mg/mL MTT solution to the well containing 90 µL of cell culture media).
2. Incubate for 2–4 h at 37 °C in the dark until purple precipitate becomes visible.
3. After incubation, remove the culture medium with excess MTT salt.
4. Add 100 µL of MTT solvent (10 mM HCl, 10% Triton-X100 in isopropanol).
5. Shake the plate for 20 min at room temperature to ensure complete dissolving of formazan crystals.
6. Measure the absorbance at 570 nm using a microplate spectrophotometer, or transfer the content of each well into an individual cuvette for separate measurements.

┌─ Take-Home Message/Lessons Learned ─────────────────────────

After reading this chapter and performing the requested assignments and exercises, students should be able to:
- Understand the difference between viability assays and functional assessment of cells and tissues
- Familiarize with most common ways to evaluate cell viability and differences between them
- Design cell viability experiment, which includes blank and positive and negative controls
- Make sample dilutions so measurements are taken within linear part of the standard curve

Self-Check Questions

Q.5.1. Which of the following assays is considered to be a tissue-specific *functional* assay?
A. Release of lactate dehydrogenase
B. Permeability to Trypan blue
C. Chondroitin production
D. Reduction of resazurin

Q.5.2. Which of the following assays is often used to noninvasively evaluate the survival of implanted tissue constructs?
A. Tensile strength
B. Load-bearing
C. Bioluminescence
D. LIVE/DEAD assay

Q.5.3. What type of control is NOT included when testing response to a drug or a treatment?
A. Blank
B. Neutral control
C. Positive control
D. Negative control

Q.5.4. Which assay recommends the transfer of accumulated media to a new well for viability testing?
A. Resazurin
B. LDH
C. MTT
D. Ethidium bromide

Q.5.5. Viability assays can rely on different aspects of cell metabolism, EXCEPT
A. Mitochondrial activity
B. Protein turnover
C. ATP production
D. DNA mutation rate

References and Further Reading

1. T.L. Riss, R.A Moravec, A.L. Niles et al. Cell Viability Assays. 2016. In: Sittampalam GS, Grossman A, Brimacombe K, et al., editors. Assay Guidance Manual [Internet]. Bethesda (MD): Eli Lilly & Company and the National Center for Advancing Translational Sciences, https://www.ncbi.nlm.nih.gov/books/NBK144065; Präbst K, Engelhardt H, Ringgeler S, Hübner H. Basic Colorimetric Proliferation Assays: MTT, WST, and Resazurin. Methods Mol Biol. (1601), 1–17 (2017)

5

Culturing Cells in 2D and 3D

Astghik Karapetyan and Narine Sarvazyan

Contents

© Springer Nature Switzerland AG 2020
N. Sarvazyan (ed.), *Tissue Engineering*, Learning Materials in Biosciences,
https://doi.org/10.1007/978-3-030-39698-5_6

What You will Learn in This Chapter and Associated Exercises
Students will learn about main types of cell cultures and the key ingredients of cell culture media. The concepts of osmolarity, pH, and sterility will be reviewed. The practical exercises will help students to gain basic knowledge of the main steps involved in media preparation, cell defrosting, and passaging.

6.1 Cell Culture

Cell culture is a combination of techniques, which enables cell isolation from an organism and subsequent maintenance of the cells in a favorable in vitro environment. There are two main ways to obtain isolated cells:

- Cells can be removed from a fresh animal or plant tissue by disaggregation that can be either enzymatic, mechanical, or both. These processes lead to disruption of the extracellular matrix, which holds cells together.
- Cells can be derived from previously established cell lines.

Primary cell cultures contain cells isolated from fresh human or animal tissue. *Passaging,* or *subculturing*, means detaching cells and transferring a small amount of them to a new flask or Petri dish with fresh culture media to facilitate further growth. *Secondary culture* is a term usually used to describe passaged primary cells. *Cell line* consists of cells, which have been continually passaged for a long time while preserving homogeneous genotypic and phenotypic characteristics. Cell lines can be *finite* and *continuous*. Cells of finite cell lines can divide, but only a limited number of times. Continuous cell lines can propagate indefinitely and are also called immortal, with immortality induced chemically, virally, or spontaneously.

There are two main types of cell cultures: adherent and suspension. For animal and human cells, suspension cultures involve mostly blood cells, although some other cell types derived from tumors can be also cultured in suspension. Adherent cells require attachment for growth and are called anchorage-dependent cells. When adherent cell cultures reach confluency (i.e., attached cells occupy most of the available surface—see ◘ Fig. 6.1), they need to be passaged. Below we will mostly discuss protocols used for adherent monolayer cultures. Most of them are also applicable to 3D cultures discussed at the end of this chapter.

6.2 Cell Culture Media

Media are a critical component of cell culture because they provide cells with the necessary nutrients, growth factors, and hormones. The basic constituents of the media include inorganic salts, amino acids, carbohydrates, vitamins, fatty acids, lipids, proteins, peptides, and trace elements. Media components also help to maintain pH and proper osmolarity of the cell environment. Different media formulations exist, each optimized for different types of cells or conditions. To chemically defined media formulations, a small amount of animal serum or chicken embryo extract is usually added to provide additional growth factors and hormones. The most commonly used serum is fetal bovine serum (FBS). Other types of serum are available,

Culturing Cells in 2D and 3D

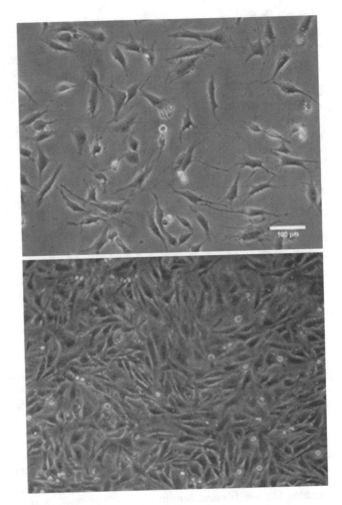

☐ **Fig. 6.1** Illustrative images of sparse (top) versus confluent (bottom) cell layers from primary culture of isolated bovine chondrocytes

including newborn calf serum and horse serum. Serum is a rich source of hormones and nutrients and is also able to bind and neutralize toxins. Disadvantages of serum include its high cost, batch-to-batch variations, and the fact that its composition cannot be precisely defined.

Three main types of cell culture media include basic media, reduced-serum media, and serum-free media. It is important for students to understand the key components of each media type as this can be critical for the success of their experiment. For example, it is perfectly fine to keep the sample in PBS for a short period of time, especially when they need to quickly check of cell morphology or behavior under the microscope. But if the goal is to observe cell contractions, then this medium will not work since it does not have physiological levels of extracellular calcium. On the other hand, having calcium in the media will significantly impair the digestion efficiency of collagenase. This is because lowering calcium concentration helps to dissociate cells by affecting adhesion molecules. A chart that summarizes the key ingredients in the main media types is shown in ☐ Table 6.1. More detailed information about different media types can be found online or in [1].

Table 6.1 Preferential use of basic salt solutions and cell culture media

Name	Key ingredients	Can be used for		
		Washing	Live cell imaging	Cell culture
Saline	NaCl	YES	NO	NO
PBS	NaCl + phosphate buffer	YES	NO	NO
Tyrode solution	NaCl, KCl, MgCl$_2$, CaCl$_2$, glucose, and phosphate	YES	YES	NO
Media	Essential salts, glucose, vitamins, amino acids, bicarbonate	Depends	NO[a]	YES
Media + serum	Media + proteins, hormones and growth factors	NO	NO	YES

[a]Unless w/o phenol red and pH adjusted with HEPES for CO_2 levels outside incubator

Basic cell culture media contain inorganic salts, amino acids, vitamins, and essential carbohydrates such as glucose. Following are the most commonly used basic medium formulations: MEM (minimum essential medium or Eagle's minimal essential medium), DMEM (Dulbecco's modified Eagle's medium), IMDM (the basic DMEM modified by Iscove), RPMI 1640 (Roswell Park Memorial Institute medium), McCoy's 5A, Opti-MEM medium, 199/190 medium, and HamF10/HamF12. Their exact chemical composition can be found on the manufacturer's website.

Reduced-serum media consists of the same basic media, which is enriched with nutrients and growth factors. The latter reduces the amount of serum to be added to basic media. Components of serum supplements usually include attachment factors such as fibronectin and laminin, growth hormones such as somatomedin, enzyme inhibitors, binding proteins, translocators, and trace elements.

Besides cell culture media, there are also washing and digestion media that are commonly used to release and wash cells during passaging:

Washing media   Balanced salt solution (BSS) is a basic medium, which is used for washing tissues and cells. It contains salts of main physiological ions such as potassium, calcium, magnesium, chloride, and sodium. Commonly used BSS names in cell culture include the following:
- Hanks' balanced salt solution (HBSS)
- Dulbecco's phosphate buffered saline (PBS)
- TRIS-buffered saline (TBS)
- Tyrode's balanced salt solution (TBSS)

Digestion media   For digestion of tissues, the proteolytic enzymes such as trypsin and collagenase are used to digest proteins of the extracellular matrix. Chelating agents such as EDTA or EGTA are commonly included to bind calcium ions that are involved in cell-cell adhesion.

6.3 Osmolarity

Osmolarity is a colligative property in that it depends on the concentration of solute ions, but not on their respective identities. Solution osmolarity is expressed as Osm/L. The unit of osmolarity is osmole, which defines the number of moles of solutes that provides an osmotic pressure of the solution. For example, NaCl dissociates into Na^+ and Cl^- ions. This means that 1 mol NaCl becomes 2 osmoles. Therefore, the solution of 1 mol/L NaCl has osmolarity 2 Osm/L. In cell culturing, osmotic pressure of media is an important factor that can affect cell proliferation. In vivo osmolarity of extracellular space is one of the most tightly controlled physiological variables unless there are changes caused by aging or disease. The osmotic pressure of human plasma is about 290 mOsm/kg, which is considered to be optimal for human cell culture. It is highly advised to have an osmometer in a cell culture lab to measure the osmolarity of media. Alternatively, it can be estimated based on the concentration of individual media components.

6.4 Sterile Environment

One of the most important procedures in cell culturing is to keep cells from contamination with microorganisms, bacteria, viruses, or fungi. These may be done by sterilizing work surfaces, media, equipment, culture waste, as well as by keeping good personal hygiene. Sterilization may be implemented with heat, filtration, UV light, as well as 70% ethanol. Air surrounding cells and their media are kept clean by using a laminar hood or other types of cell culture hood. The cell culture hood is an enclosure into which sterile air is forced through a HEPA-like filter.

An essential piece of equipment, called an autoclave, sterilizes instruments and glass bottles used to prepare and store media. It uses pressurized steam to heat the material to be sterilized. Autoclaving effectively kills most microbes, spores, and viruses.

Filtration is used for media that contain thermally labile components. It employs filters with pores so small that microorganisms can't pass it. Most of the filters used for sterilization are made from porous cellulose acetate and have either 0.22 µm or 0.44 µm pores.

Below are some basic rules that can help to avoid cell culture contamination:

— The working surface and surrounding areas, as well as equipment, must be disinfected before and after each use. Ultraviolet (UV) light can be used for sterilization when left on for a significant amount of time. In order to prevent damage to skin and eyes, UV light must be turned OFF when actual work in cell culture hood is being done.

— Before and during the work, a person is required to wipe work surface and hands with 70% ethanol. Particular care has to be exerted in cases of any media spillage. Everything, including plates, dishes, pipettes, has to be wiped with ethanol before entering the workplace inside the hood.

— Media, reagents, solutions, plates, dishes, pipettes, tips, and flasks have to be sterilized.

— Avoid pouring reagents or solutions directly from the bottle. Instead, use pipettes to draw needed amounts.

- Don't leave the solutions uncovered in an open environment.
- Put the cap of bottles on the surface with the cap opening facing down.
- Wash hands before and after work.
- Wear personal protective equipment (PPE) such as gloves, laboratory coat, safety visor, overshoes, and head cap.
- Don't converse or sing during the preparation of solutions.
- Do the work as quickly as possible in order to avoid contamination.

Antibiotics are often added to the cell culture media in order to prevent contamination. However, their long-term usage can cause the formation of antibiotic-resistant cell strains. In addition, antibiotics can interfere with differentiation protocols or maintenance of stem cell cultures. Hence, for such cases, the use of antibiotics should be avoided or minimized as much as possible. On the other hand, most protocols to isolate cells for primary cell culture are not done in sterile conditions. Therefore, primary cell cultures almost always require the use of an antibiotic. The most common antibiotics used in cell culture are ampicillin, penicillin, streptomycin, and gentamicin.

6.5 Temperature

As mentioned above, for the optimal growth of most of the mammalian cells, a temperature of 36–37 °C is needed. For short-term handling of the cells, including washing, re-plating, imaging, and dye loading, room temperatures are perfectly suitable. Placing cells in a refrigerator for a short term (10–30 min) is not detrimental. Overheating cells to temperatures above 50 °C leads to irreversible cell damage. If cells are frozen at temperatures below −4 °C without specific precautions, ice crystals formed within the cells rupture cell membrane leading to cell death.

6.6 Concept of pH

pH is one of the most important physicochemical factors for maintaining cell growth. pH is a measure of hydrogen ion concentration in solution and used to specify acidity and basicity of the solution under appropriate temperature. The equation, which defines pH, is: $pH = -\log [H^+]$, where $[H^+]$ is a concentration of hydrogen ions. For example, if the concentration of hydrogen ions in solution is 0.00001 M, the pH of the solution will be equal to 5.

$$pH = -\log[0.00001] = -\log\left[1 \times 10^{-5}\right] = -(-5) = 5$$

pH value can range from 0 to 14. Solutions with a high concentration of hydrogen ions have a low value of pH, and solutions with a low concentration of hydrogen ions have a high value of pH. Acidic solutions have a pH lower than 7, while basic solutions have a pH higher than 7. Solutions with a pH of 7 are considered neutral.

The physiological pH falls within a narrow range of 7.2–7.5. The most commonly used buffer for maintaining a stable pH of solutions outside of CO_2 incubator is

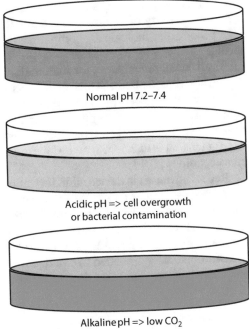

Visual changes in the color of cell culture medium containing phenol red allow making conclusions about pH

HEPES (4-(2-hydroxyethyl)-1-piperazineethanesulfonic acid). It can be made as a 0.5M–1M stock, pH of which is brought to a range of 7.2–7.4 using concentrated HCl solution. It can be then added to the media to achieve the final 10–20 mM concentration.

Any cell culture lab must own a sensitive pH-meter to ensure pH measurements of prepared media. Many media formulations include an indicator called phenol red. It allows a user to constantly visually monitor the pH status of the media by its color (■ Fig. 6.2). If the medium color turns yellow (acidic pH), it is likely an indicator of cell overgrowth or bacterial contamination. If the medium color turns purple (alkaline pH), this usually indicates a lack of CO_2. The latter can be a result of incubator dysfunction or due to simply having a cell culture dish for too long.

6.7 Cell Culture Incubator

One of the required pieces of equipment for cell culturing is cell incubator, which provides an appropriate environment for cell growth. It must have forced-air/CO_2 circulation and temperature control within ±0.2 °C. Three main environmental conditions provided by cell incubators include physiological temperature of 36–37 °C, 85–95% humidity, and 5% carbon dioxide. These conditions mimic the environment to which cells are exposed while inside the body. High amount of gaseous carbon dioxide within the cell incubator leads to acidification of the media due to the conversion of CO_2, once it is dissolved, into carbonic acid. To counteract this effect, most media formulations require the addition of 2 g/L sodium bicarbonate. To prevent significant pH changes due to taking cells in and out of the incubator as well as metabolic activity of the cells that leads to CO_2 production, the above-mentioned HEPES or other biologically compatible buffers can be also included in cell culture media.

Traditionally incubators for mammalian cell cultures are connected to 95% air / 5% carbon dioxide tanks. Thus, inside them, the concentration of oxygen is close to its atmospheric concentration of ~20%. This is equivalent to 150 mmHg partial pressure of oxygen (760 mmHg atmospheric pressure multiplied by 20%). Since it has been shown that many cell types, particularly stem cells, are better maintained when oxygen tension is at 2.5–5%, many labs are switching to gas tanks with lower oxygen concentrations [2].

Recent advances in device miniaturization enable visual monitoring of cells inside incubator via wireless access. An example of such a device is CytoSmart Lux2. It is a compact inverted microscope for brightfield live-cell imaging. The device can fit inside a standard incubator and can be programmed to acquire and send to the user's computer magnified pictures of cell culture at specified time points. The images are then automatically uploaded to cloud storage where they can be analyzed for confluency, cell division, differentiation, or other morphological changes.

6.8 Cell Storage, Defrosting, and Passaging

Since continuous cell lines are passaged many times, the risk of their genetic instability increases. Therefore, it is very important to keep the stock of these cells in cryogenic storage. Freezing temperatures are not suitable for this purpose, because at regular freezer temperatures (from −20 to −80 °C) the cell viability decreases. For long-term storage of frozen cells, Dewar containers filled with liquid nitrogen are used. They enable maintaining temperatures of −196 °C.

Defrosting cryopreserved cells without significant loss of viability requires quick and careful work. Below are detailed steps for successful thawing (Fig. 6.3):

- Remove the cryovial containing the frozen cells from liquid nitrogen storage and immediately place it into a 37 °C water bath. It is recommended to cautiously dip the vial in the water bath without exposing the threads or the top lip of the vial to the water to prevent contamination.
- Quickly thaw the cells (< 1 min) by gently swirling the vial in the 37 °C water bath until there is just a small bit of ice left in the vial.
- Transfer the vial into a laminar flow hood. Before opening, wipe the outside of the vial with 70% ethanol.
- Transfer the thawed cells dropwise into the centrifuge tube containing the desired amount of pre-warmed complete growth medium appropriate for the selected cell line.
- Centrifuge the cell suspension at approximately 200 g for 5–10 min. The actual centrifugation speed and duration vary depending on the cell type.
- After the centrifugation, check the clarity of supernatant and the visibility of a complete pellet. Aseptically decant the supernatant without disturbing the cell pellet.
- Gently resuspend the cells in complete growth medium and transfer them into the appropriate culture vessel and into the recommended culture environment.

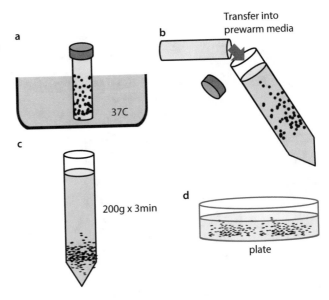

Fig. 6.3 Steps to cryopreserve cells. **a** Warm sample in 37 °C waterbath, **b** transfer cells into a centrifuge tube, **c** centrifuge at 200 g for 3–5 min, **d** plate cells

Note: The appropriate flask size depends on the number of cells frozen in the cryovial, and the culture environment varies based on the cell and media type.

After cells are defrosted and cultured for several days, they reach a state of confluency. For adherent cultures, this means that cells occupy most of the available surfaces. Afterward, the proliferation of the cells dramatically decreases. In order to keep cells alive and growing, they must be split (this process is also called passaging or subculturing). Techniques for cell passaging differ depending on the cell type. Most commonly, it involves the addition of media containing trypsin to detach cells, followed by cell centrifugation, media replacement, and cell count. Afterward, cells are plated using an attachment surface that is about 5–10 times larger than the original surface (**Fig. 6.4**).

The content of the above sections is a compilation of useful information about 2D cell culture techniques from multiple freely available sources [3–5]. Students are strongly encouraged to read the full text of these articles.

6.9 3D Cultures

Cells grown in a 3D environment have different morphology compared to those grown in 2D on plastic or glass surfaces. They have also shown to differentiate and be affected by the drugs differently. Therefore, an increasing number of researchers are conducting experiments using 3D culture approaches. There are three main ways by which this is currently done (**Fig. 6.5**).

The first one is an *aggregation* of cells to form 3D spheroids, which can be *gravity- or stirring based*. When stem cells are used as starting material, these spheroids are called embryoid bodies (abbreviated as EBs). 3D cell spheroids can be created by using a "hanging drop" approach or by seeding cells in special cell culture plates with multiple miniwells having a round or conical bottom (an example being Sphericalplate 5D or similar designs). The hanging drop approach is laborious, but it does not

6

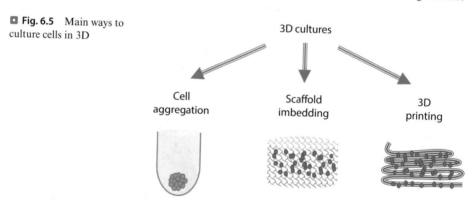

a Remove media **b** Trypsin-EDTA **c** Fresh media

d Fresh media **e** Multiple plates

200g × 3min

☐ **Fig. 6.4** Subculture procedure. **a** Remove media, **b** incubate with Trypsin-EDTA, **c** neutralize by adding fresh media, **d** collect and centrifuge, **e** resuspend in fresh media and plate using 5–10 times larger surface

☐ **Fig. 6.5** Main ways to culture cells in 3D

3D cultures

Cell aggregation   Scaffold imbedding   3D printing

require any specialized equipment as droplets of solutions containing anywhere from 200 to 400 cells are simply pipetted on the lid of the standard cell culture place and then flipped during subsequent culturing. The miniwell approach enables upscaling the number of spheroids to thousands in a single plate, improves their homogeneity, and makes it easy to perform media exchange. The formed spheroids can then be even used as individual blocks for 3D printing applications. 3D cell aggregates can also be formed when certain types of suspended cells are stirred within cell culture flask for extended periods. The shape and size of such aggregates are more variable than 3D spheroids formed by hanging drop or miniwell approaches (☐ Fig. 6.6).

The second way to culture cells in 3D is to seed on them on different types of meshes and/or to combine such scaffold with hydrogel embedding. All the criteria— fiber diameter, orientation, porosity, degradation rate, and mechanical properties— can be adjusted to create the most suitable 3D culture platform. Such meshes can be 3D printed using PCL polycaprolactone, electrospun from PLLA (poly(L-lactide))

□ **Fig. 6.6** Main methods for self-assembly of 3D cell spheroids

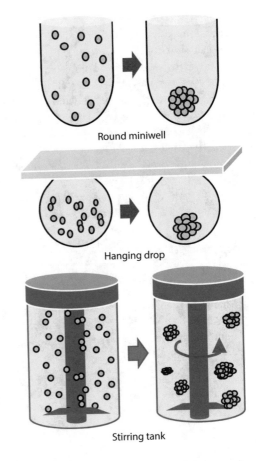

Round miniwell

Hanging drop

Stirring tank

or other polymers. Numerous companies including Bellamesh, Mimetix, The Electrospinning Company, and many others can make scaffolds based on user's needs. Different mesh sizes and structures have been shown to dramatically affect cell behavior and morphology so screening of several mesh types is needed to optimize a particular mesh to the application of an individual user. 3D cancer models, cartilage and bone engineering, and vascularization studies can all benefit from using 3D versus 2D cell culture systems.

The third way to culture cells in 3D is to use *bioprinting*, which will be covered in more detail in ▶ Chap. 10.

When 3D tissues are small (less than 200 microns), they can be kept in cell culture dishes and cultured similarly to 2D cultures. For larger 3D constructs, it is better to switch to perfusion-based culturing or use bioincubators (discussed in ▶ Chap. 11). Doing so greatly increases the access of the cells within such constructs to the nutrients. And since combining a pump and a cell incubator is often a cumbersome endeavor, new devices such as TEB500 Cell Culture Bioreactor recently came to the market. The latter has a functionality of a CO_2–O_2 incubator but also incorporates an integrated double peristaltic pumping system (EBERS Medical, Zaragoza, Spain).

Session I

Demonstration

The instructor demonstrates media preparation, passaging, counting, and storing cells using a commercial cell line. Alternatively, fibroblast or endothelial cells obtained during the week 4 demo can be kept in culture and used for these experiments. The concept of 3D culture by making spheroids using hanging drop can also be shown and discussed.

Homework

Each team selects from literature three different protocols that can be used to culture cells from their organ of choice. A table comparing major differences and similarities in these three culturing conditions is compiled to be discussed in the class.

Session II

Team Exercises

Each team is given one plate of cells to passage using their own media. The latter has to be prepared from media powder to which appropriate amounts of bicarbonate have to be added. This has to be followed by filtering and the addition of serum and antibiotics. Treated coverslips from week 3 can be used to plate excess cells on different adhesion surfaces.

Homework

Each team prepares a detailed protocol that describes steps involved in making culture media, passaging the cells, and monitoring their proliferation on different adhesion surfaces.

ℹ️ Sample Protocol

Cell passaging

1. Warm to room temperature all necessary solutions including trypsin, culture media, and PBS (phosphate-buffered saline).

2. Under the laminar hood, carefully remove media from cell plate using sterile pipette tips connected to a pipettor or a vacuum line.

3. Add a small volume of PBS to the dish, gently swirl it around, and carefully remove it using pipette or aspirator.

4. Add 0.05% of trypsin (the volume of added trypsin depends on the plate surface) and place the cell plate into the incubator for about 1 min. Place cell plate under a phase-contrast microscope and check cell appearance. If they look round, then they have started to detach from the plate surface. It is important to visually confirm that cells are properly lifted from the plate before neutralization.

5. To neutralize the impact of trypsin, add the same volume of media. Resuspend the cells in the media and transfer the cell suspension into a 15 or 50 mL tube. To calculate the number of cells, mix 10 μL of cell suspension with the same volume of Trypan blue dye and transfer it to the hemocytometer for cell count (see ▶ Sect. 4.5 of ▶ Chap. 4).

6. Centrifuge cell suspension for 5 min at 1500 rpm.
7. Remove the supernatant and carefully break the pellet by gentle triturating (pipetting back and forth) to separate cells from each other.
8. Add more media and carefully pipette cell suspension into a new plate with growth media (the volume of cell suspension depends on the number of cells). Then place the plate into an incubator for 24 h, followed by daily or bi-daily media change.

Take-Home Message/Lessons Learned

After reading this chapter and performing the requested assignments and exercises, students should be able to:
- Understand differences between different cell culture types
- Make cell culture media from commercial powder formulations
- Adjust media pH, check its osmolarity, and sterilize it
- Perform basic steps to passage cells in sterile conditions
- Understand the main principles to culture cells in 3D

Self-Check Questions

? Q.6.1. Cell cultures cannot be classified as either _____
    A. Suspension or adherent
    B. Finite or continuous
    C. Plant or animal
    D. Synthetic or natural

? Q.6.2. Ion present in ALL types of cell culture and cell washing media is
    A. Potassium
    B. Sodium
    C. Calcium
    D. Magnesium

? Q.6.3. Cell culture media typically includes all these components, EXCEPT
    A. Buffer
    B. Nutrients
    C. Antibodies
    D. Antibiotics

? Q.6.4. In typical cell culture, this is NOT a controlled parameter
    A. Temperature
    B. Carbon dioxide levels
    C. Humidity
    D. Nitrogen levels

Q.6.5. Choose the correct statement about the pH of culture media.
  A. pH is not affected by levels of carbon dioxide within an incubator.
  B. pH is affected by incubator humidity.
  C. pH must be between 7.2 and 7.4 pH units.
  D. pH is not affected by the presence of a buffer.

References and Further Reading

1. M. Arora, Cell culture media: A review. Mater Methods **3**, 175 (2013)
2. C. Mas-Bargues et al., Relevance of oxygen concentration in stem cell culture for regenerative medicine. Int. J. Mol. Sci. **20**(5) (2019)
3. Z. Yang, H.-R. Xiong, Culture conditions and types of growth media for mammalian cells, in *Biomedical Tissue Culture*, IntechOpen books, (2012). https://doi.org/10.5772/52301.
4. Invitrogen® and Gibco™, *Cell Culture Basics Companion Handbook* (Thermo Fisher Sci. Inc.). Waltham, Massachusetts, United States (2014)
5. Sigma®, Fundamental techniques in cell culture laboratory handbook, in *European Collection of Cell Cultures*. St. Louis, Missouri, United States (2015)

Imaging, Staining, and Markers

Vardan Avetisyan and Narine Sarvazyan

Contents

© Springer Nature Switzerland AG 2020
N. Sarvazyan (ed.), *Tissue Engineering*, Learning Materials in Biosciences,
https://doi.org/10.1007/978-3-030-39698-5_7

What You will Learn in This Chapter and Associated Exercises

The goal of this chapter is to cover the key concepts related to imaging, fluorescence, and sample labeling. Students will gain basic knowledge of viewing cells and tissues using a microscope as well as how to stain different structures using specific markers.

7.1 Visible Light

Humans observe the world using what is appropriately called the *visible light spectrum*. It corresponds to roughly 400–750 nm range of wavelengths. It is neighbored by the ultraviolet (UV, 10–400 nm) and infrared (IR, from 750 nm until about 1000 microns) light ranges. When photons of certain energy are absorbed by a specimen, it can be measured using spectrophotometer or seen as a darker area on microscopic images. This phenomenon is commonly referred to as *light absorption*.

7.2 Fluorescence

Fluorescence is the ability of molecules to emit light after absorbing photons of higher energy. The energy coming from a light source shifts an electron of an atom from a lower energy state to a higher (also called "excited") energy state (◻ Fig. 7.1). When fluorophore returns to its ground state, energy is released in the form of another photon. This newly emitted photon has lower energy and therefore lower frequency. Having a lower frequency (ν) means that the photon has a longer wavelength (λ). For example, if fluorophore emits red light, the sample needs to be illuminated using

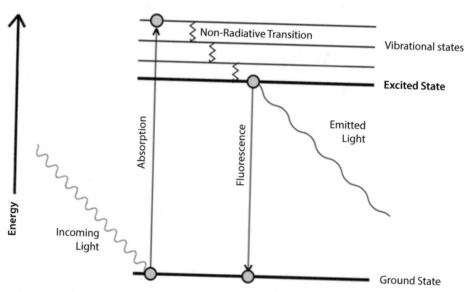

◻ **Fig. 7.1** Jablonski diagram. This diagram illustrates the energy states of molecules and the transitions between them. Due to energy losses in an excited state, emitted photons always have lower energy than absorbed photons. Therefore, the emission wavelength is always longer than the excitation wavelength

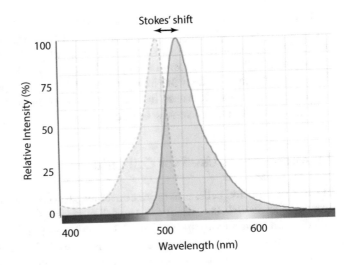

Fig. 7.2 Example of excitation (shaded blue) and emission (shaded pink) spectra of Fluo-4, an indicator for intracellular calcium levels

higher energy wavelength (which will appear as green or blue light). The distance between optimal excitation wavelength and wavelength that has maximal emission intensity is called Stokes' shift. The larger the Stokes' shift is for a specific dye, the easier it is to select appropriate filters to image samples stained with it (◘ Fig. 7.2).

Today, a great number of fluorescent dyes have been developed. These dyes are capable of selectively labeling intracellular components of both live and fixed cells, to monitor the release of ions, to track the movement of cells or proteins, to examine the oxidative state of the cells, and the list goes on [1]. Many of these dyes are added to the cells or tissue in AM (*acetoxymethyl ester*) or DA (*diacetate*) form, which makes it easier for the dye to pass through cell membranes. Upon entering the cells, these added chemical groups are cleaved off by intracellular enzymes, making the active form of the dye trapped inside the cell.

In addition to low molecular weight dyes mentioned above, there is also a wide range of fluorescent proteins that were developed after the discovery of green fluorescent protein (GFP) [2]. Genes for GFP or its many analogs can be included in the DNA of an organism or individual cells. This results in cells being permanently autofluorescent, which is very helpful for tracing cells that are imbedded into scaffolds for extended periods. The same applies to tracing cells upon their implantation into live tissues.

7.3 Basics Light Microscope Types

All microscopes have a goal of creating a magnified image of a specimen. They differ in the way the image is obtained, how the specimen is placed relative to the objective, the degree of magnification, spatial resolution, type of sensors, and other features. *Light Microscope* uses lenses and visible light to create a magnified image of the object [3]. There are two main types of light microscopes: upright and inverted. In case of an upright microscope, the object to be observed is positioned above the objective, while in case of an inverted microscope, objectives are placed beneath the object. The light from a light source on an opposite side passes through the object to the objective lens. Common magnification values include 4×, 10×, 20×, or 100×. The light then passes through the second pair of lenses located within eyepieces. The latter usually magnifies the image for an additional ten times. So, the user looking at let's say 20× objective using 10× eyepiece lenses ends up seeing 200 times larger object. *Stereo microscope* uses two separate optical paths from the objective lens to eyepieces. The different angles of these two pathways create a 3D image of an object. This type of microscope usually has low magnification. It is used to study the surfaces of solid structures, organ dissection, cannulation of small vessels, or microsurgery.

7.4 Microscopes Used for Cell Culture

In cases of cultured cells, it is more convenient to observe samples from beneath in order not to disturb the sterility of the cell environment. This is enabled by what is called an *inverted microscope* (◪ Fig. 7.3). In the case of inverted microscopes, light is coming from the top, while the objective is beneath the stage. Another feature that is useful to observe cells is called phase contrast. It enables to enhance the contrast of transparent and colorless objects such as cells or thin tissue slices. To have this feature, the microscope needs to have a special plate that is inserted into its light path. The phase contrast plate must match with the objective. The difference in phase of the light wave is not noticeable to the human eye; however, when the change in phase is increased to half a wavelength by phase-plate, it causes a visible difference in brightness enabling to better see cell boundaries and other structures.

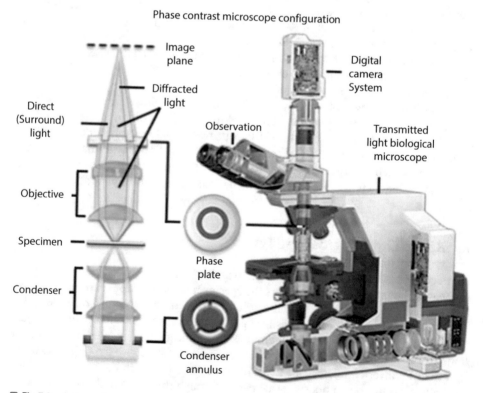

◪ **Fig.7.3** A general appearance of an inverted, phase contrast cell culture microscope

7.5 **Fluorescent Microscopes**

Fluorescent microscopes employ a set of filters that enables separation of illuminating and collected photons based on their wavelength. To do so, the light path includes an excitation filter, a dichroic mirror, and an emission filter (◻ Fig. 7.4). The filters and the dichroic mirror must be chosen to match the spectral properties of the dyes used to label the specimen. Several different fluorescent dyes can be imaged simultaneously by selecting the appropriate combination of filters so neither excitation nor emission ranges overlap. This enables the combination of several single-color images into one.

Let's consider an example. When selecting a dye to be used with a confocal microscope, the first step is to determine what lines of the available laser can excite the dye. In the case of Fluo-4 dye (spectra of which are shown in ◻ Fig. 7.2), 488 nm argon laser will be a good fit as far Fluo-4 excitation wavelength. The next step is to select a dichroic filter that can separate the two peaks (something around 500 nm will fit). Lastly, we need to select an appropriate emission filter (any filter that lets photons from 520 to 560 nm to pass through should work).

Filters that pass wavelengths *within a certain specified range* are called *bandpass filters*. Filters that allow light to pass with wavelengths *longer than the specified value* are called *long-pass filters*. Filters that let photons pass with a wavelength shorter than specified values are called *short-pass filters*. By changing multiple filters or excitation wavelengths, images of samples stained with several dyes can be obtained. In those multistained samples, it is critical to eliminate what is called "optical crosstalk." The latter occurs when filters are not optimized and there is insufficient separation between photons of different wavelengths. In those cases, a signal from one dye will "bleed" into a channel reserved for another dye.

◻ **Fig. 7.4** Basic principles behind fluorescence microscopy. The light source can be a lamp, a laser, or an LED source

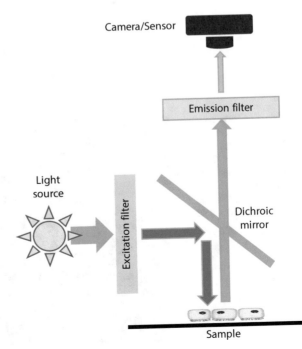

7.6 Confocal Microscopy

In regular fluorescence microscopes, photons are coming from the full thickness of the specimen; thus, it is not possible to acquire crisp images from a single focal plane. In contrast, confocal microscopy eliminates this out-of-focus information by means of a confocal "pinhole" situated in front of the image plane. It allows only the in-focus portion of the light to be collected. Light from above and below the plane of focus of the object is eliminated from the final image. A diagram of the confocal principle is shown in ◘ Fig. 7.5.

Laser scanning confocal microscopes use lasers as the light source. The laser beam is then "scanned" through the object by using x-y scanning mirrors, while altering the focus allows the acquisition of different focal planes. Multiple planes are then reconstructed into 3D images using a computer. Confocal microscopy is also widely used to acquire crisp pictures of individual stains, which can be then combined into a pseudocolor image of a sample with each cellular target assigned a different color (◘ Fig. 7.6).

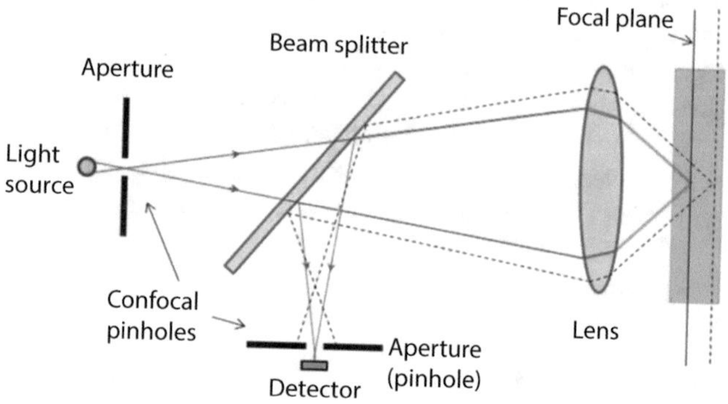

◘ **Fig. 7.5** Diagram showing the main principles behind confocal microscopy. The presence of pinholes removes out-of-focus light enabling optical sectioning of the sample

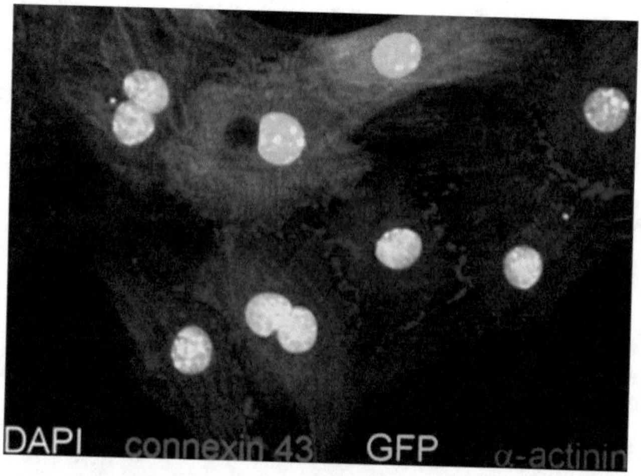

◘ **Fig. 7.6** Confocal image of cardiac myocytes derived from GFP-expressing mouse embryonic stem cells. Cells were fixed and stained with DAPI (white), and with antibodies for alpha-actinin (Cy5, red) and connexin 43 (Cy3, blue)

7.7 Electron Microscopes

In the case of electron microscopy, the light source is replaced by a stream of electrons that pass through a set of electromagnets. The wavelength of an electron is much smaller than the visible light, so this approach allows one to create much more detailed images.

Scanning electron microscope (SEM) creates a 3D image of an object with high resolution and magnification by covering the source of an object with an electron reflecting film. So, the cells or tissue structures are essentially observed "from an outside."

Transmission electron microscope (TEM) creates highly detailed 2D images of cell infrastructure as electrons pass through a very thin specimen while being absorbed by its denser components. Other types of microscopes are currently available, and interested students are referred to numerous online resources on this subject.

7.8 Histology

In addition to fluorescent dyes, many histological dyes can be very useful in TERM applications. In general, they are not as specific as fluorescent markers and are best used for tissue rather than for intracellular staining. Histology procedures are lengthy and involve sample fixation, mounting, slicing, staining, and finally imaging. Most of histology protocols can be found online [Ref: https://webpath.med.utah.edu/HISTHTML/MANUALS/MANUALS.html]. Due to the time it takes to perform all the steps, it is advised to use commercial histological facilities for better results.

7.9 Dye Aliquoting

Multiple cycles of freeze-thawing can affect the integrity of the samples, including fluorescent dyes. Therefore, the first step in any staining protocol is to divide the dye stock solution into small aliquots that can be individually defrosted for individual experiments. To perform aliquoting, one needs to estimate the most reasonable volume of individual aliquots based on the stock concentration and required dilution to achieve the volume of solution that is sufficient for one experiment. Most of the commercial dyes come as 10 mM stock solution in dimethyl sulfoxide (DMSO) or have to be diluted in DMSO per the product manual. For example, let's consider ThermoFisher nuclear stain 7-AAD (#A1310). It comes as a 1 mg solid. If diluted with either 78 µL methanol or DMSO, it will yield 10 mM concentration (7-AAD molecular weight is 1270). Considering the final dye concentration being about 5 µM, a 10 mM stock solution will require an additional 1:2000 dilution. Since 2 mL dye solution is usually enough to fully cover several glass coverslips, making 18 × 1 µL aliquots and 6 × 10 µL aliquots (for further aliquoting) makes a good sense. To make aliquots, multiple small size Eppendorf tubes have to be labeled, followed by the addition of a specified volume (1 µL or 10 µL in our case) of the dye to each of the Eppendorf tubes. The label needs to include the name of the dye, manufacturer and catalog number of the dye, the day of aliquoting, stock concentration, and aliquot volume. Aliquots are then frozen (unless product manual notes otherwise).

7.10 How to Observe Stained Samples

When imaging samples at low resolution using air objectives ranging from 2× to 40×, pretty much any clear dish can be used. Glass coverslips are preferable when looking at fluorescent samples, especially when a non-confocal setup is used to observe it. Higher resolutions usually require immersion oil (the latter is stated on the objective), in which case samples can be only imaged through 170 microns thick or what is called #1.5 size glass coverslip. When imaging fixed samples, this can be done by simply flipping samples with a coverslip surface facing the objective. Mounting media, such as Mowiol, are usually used to fix cells or thin tissue between the glass slide and the coverslip. Imaging of live cells requires medium presence; therefore, special chambers must be used. They can be bought or custom-made. Their main feature is 170-micron thick glass coverslip at the bottom of the dish that allows taking images of the sample from below (◻ Fig. 7.7).

Alternatively, for live samples special water immersion objectives can be used. These objectives use water in place of the oil as the immersion medium. They enable approaching live sample from the top since these objectives can be in direct contact with the saline that surrounds living cells or tissues. Unfortunately, it is nearly impossible to maintain sterility of these samples afterward, unless the entire imaging setup is placed in a sterile environment.

7.11 Choosing the Dyes

Online resources such as the ThermoFisher Fluorescence SpectraViewer website can be used to find emission and excitation spectra for selected dyes. Multiple dyes can be added and removed from the graph, enabling to see overlaps in excitation and emission spectra. It is important to select dyes that can be excited with light sources available to a particular lab or the user. For example, if a user's confocal system does not have a UV laser, then he/she will not be able to use DAPI as its excitation lies in the 300–400 nm range.

◻ **Fig. 7.7** Sample placement for an inverted microscope with oil immersion objectives. Typical 60× or 100× oil objectives require 170-micron thick glass coverslip between the sample and the objective lens

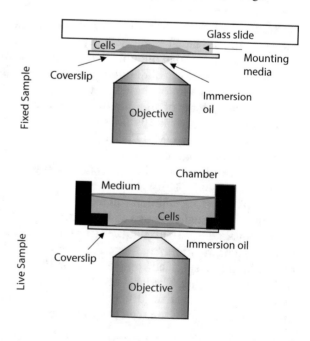

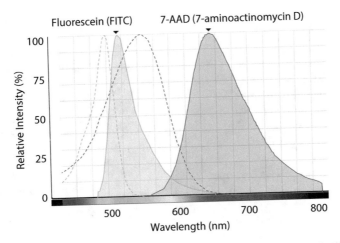

□ Fig. 7.8 Principle of selecting suitable dyes for staining with multiple markers. Excitation spectra of FITC (dotted green line) indicate that it can be excited by blue light (about 490 nm) while collecting emitted green light at 510–530 nm range. Samples can be co-stained with 7-AAD, a red nuclear stain, that has excitation and emission peaks at 560 and 650 nm, respectively. Based on the spectra, the expected overlap between the two stainings should be negligible

It is recommended to choose dyes with well-separated emission or excitation spectra. For example, let's say the user wants to stain cells with 7-AAD (for nuclei) and TRITC-conjugated phalloidin (for actin filaments). Both can be excited by a 535 nm laser. Emission spectra are different, but there is a significant overlap in almost all areas of the spectrum. It is possible to use narrow emission bands to separate signals from these two fluorophores (i.e., 580–600 for TRITC and 660–740 nm for 7-AAD), but this will translate into low-intensity signals. Instead, the user will be better off by replacing 7-AAD with a different nuclear stain—let's say TO-PRO-3 or substitute TRITC-labeled phalloidin with FITC-conjugated phalloidin (□ Fig. 7.8). In both the cases, there will be a clear separation between peaks of both excitation and emission spectra.

Session I
Demonstration
The instructor shows how to use different light paths on available fluorescent scopes to perform imaging of samples stained with different dyes. If a confocal setup is available, the use of Z-stacks to image tissue slices is also shown. Lastly, students are shown how to fix and stain cultured cells on a glass coverslip using multiple organelle-specific dyes. Previously stained samples with clearly delineated cell morphology (similar to the one shown in □ Fig. 7.6) can be used to demonstrate different optical paths and how specific filter can affect the intensity of the individual channels.

Homework
Students are asked to familiarize themselves with the spectral properties of a given list of dyes to be used during Session II.

Session II

Team Exercises

Using known dye spectra and a list of available probes, each team selects the best dye combinations and filter settings to perform triple staining of given cell samples using DAPI, phalloidin, mitochondrial, or other organelle-specific markers. For this exercise, students need to match available dyes with hardware settings of the confocal system or any other lab-specific fluorescence imaging equipment. Cells isolated and plated by the teams during previous weeks can be used for these experiments.

Homework

Students are tasked with taking images of their stained samples and making PowerPoint presentations that show respective cell structures.

ℹ Sample Protocols

Cell preparation for team exercises

1. Plate cells on gelatin-coated coverslips.
2. Culture to desired density/age.
3. Wash two times with PBS (phosphate-buffered saline).
4. Cover cells with 4% formaldehyde for 10 min at room temperature. Alternatively, cover cells with ice-cold 1:1 methanol-acetone solution and put the cell plate in the freezer for 10 min.
5. Wash two times with PBS.
6. Add 400 μL Wash Buffer/Blocking Buffer (PBS + 0.1% serum/1% albumin + 0.02–0.03% sodium azide).
7. Stain or store at 2–8 °C for up to 3 months.

Multi-dye staining

1. Choose dyes according to chosen organelles (e.g., nucleus, cell membrane, mitochondria, etc.). Make aliquots. Use one aliquot to make a dye solution with recommended final concentration.
2. Transfer fixed cells grown on 170-micron thick glass coverslips to a small, dry Petri dish. Coverslips of different sizes can be used depending on the final observation chamber.
3. Carefully add 10–20 μL of dye, diluted to a final concentration, on top of each coverslip. Cover Petri dish with a lid and wait 15 min at room temperature.
4. Wash coverslips two times with PBS (phosphate-buffered saline).
5. Check for positive cell staining using a fluorescent microscope with filters setting suitable for the dye. This can be done using low magnification objectives that do not require close contact with the sample.
6. Add second dye and then third dye as per steps 3–5.
7. Use high-resolution objectives to obtain good quality images illustrating different cell structures.

Mounting slides with Mowiol. Mowiol is a proprietary solution of polyvinyl alcohol. It fully hardens overnight and does not require sealing coverslips with nail polish. The latter is another way to mount samples. Opened aliquots of Mowiol can be stored in the fridge for about 1 month.

1. Wash coverslip with saline.
2. Put a small drop of Mowiol (5–10 µL) on a glass slide. Make sure that there aren't any bubbles.
3. Take coverslip with tweezers, hold it with cell surface down, and put on Mowiol drop while tilted.
4. Label the glass slide.
5. Wait at least 30 min for air-objective and 24 h if oil objectives are to be used for imaging. It is easy to smear Mowiol while it is not fully settled. Therefore, oil objectives, which come to very close contact with the surface of the coverslip, can shift the coverglass leading to irreversible sample damage.

Take-Home Message/Lessons Learned

After reading this chapter and performing the requested assignments and exercises, students should:

- Know the basic functions of different types of microscopes and objectives.
- Be able to select appropriate type of filters when viewing sample stained with a specified dye(s)
- Learn how to fix and stain cell or tissue samples with different fluorescent markers
- Be able to aliquot dyes for long-term storage
- Know how to use the online spectral libraries to select multiple dyes with appropriate emission/excitation properties.

Self-Check Questions

? Q.7.1. You stained cell samples with mitochondria-specific dye and want to observe individual organelles, which are approximately 1–2 micron in size. The most appropriate objectives will be

    A. 2×
    B. 10×
    C. 20×
    D. 100×

? Q.7.2. You are trying to get a single image of a 2 × 2 mm piece of engineered tissue. The most appropriate objectives will be

    A. 2×
    B. 10×
    C. 20×
    D. 100×

? Q.7.3. Use online spectra finder to determine if a single excitation source can be used to image sample co-stained with _____ antibody and a nuclear stain ___.

    A. FITC & TO-PRO-3
    B. FITC & 7-AAD
    C. TRITC & TO-PRO-3
    D. TRITC & 7-AAD

❓ Q.7.4. Based on 7-AAD spectra, the following set of filters (excitation/dichroic/emission) should be suitable to image samples stained with it:

A. 480 nm short-pass, 500 nm, 600–650 nm band-pass

B. 540–570 nm, 600 nm, 620 nm long-pass

C. 480 nm long-pass, 650 nm, 600–650 nm band-pass

D. 500 nm long-pass, 550 nm, 600 nm short-pass

❓ Q.7.5. Choose the correct statement.

A. A confocal microscope uses a beam of electrons to create crisp images of tissue at different depths.

B. Mowiol aliquots cannot be stored and must be prepared fresh.

C. The use of acetoxymethyl ester AM form of the dye enables to store stained cells for at least a month at room temperature.

D. The surface of a sample to be observed using a 63× oil objective with ∞/0.17 marking must be less than 170 microns away from the objective lens.

References and Further Reading

1. J.T. Russell, Imaging calcium signals in vivo: a powerful tool in physiology and pharmacology. Br J Pharmacol. **163**(8),1605–1625 (2011)
2. E.A. Rodriguez, R.E. Campbell, J.Y. Lin et al. The Growing and Glowing Toolbox of Fluorescent and Photoactive Proteins. Trends Biochem Sci. **42**(2), 111–129 (2017)
3. K. Thorn, A quick guide to light microscopy in cell biology. Mol Biol Cell. **27**(2), 219–222 (2016)

Stem Cells and the Basics of Immunology

Zaruhi Karabekian and Narine Sarvazyan

Contents

© Springer Nature Switzerland AG 2020
N. Sarvazyan (ed.), *Tissue Engineering*, Learning Materials in Biosciences,
https://doi.org/10.1007/978-3-030-39698-5_8

What You will Learn in This Chapter and Associated Exercises

Students will learn about different types of stem cells and their role in tissue engineering applications, including examples of cell type–specific differentiation protocols. A brief overview of immunology as applied to implantation of engineered tissue will be given. The concept of induced pluripotency and its implication for the field will be introduced. Students will then practice immunostaining their own samples using given choice of primary and secondary antibodies.

8.1 Concept of Stem Cell Potency

Stem cells can serve as precursors to many types of specialized cells and are capable of unlimited division without losing their stemness (◻ Fig. 8.1). *Potency* specifies stem cell potential to differentiate into various cell types. Different degrees of stem cell potency exist. *Totipotent cells* are found in the very early morula stage of the embryo and are capable of forming all of the body cells as well as the placenta. *Pluripotent cells* originate from the inner cell mass of the blastocyst. They cannot form the placenta but are capable of forming all cells of the body. *Multipotent stem cells* can produce several types of cells functionally related to each other. *Oligopotent stem cells* can differentiate into only a few, usually phenotypically related, cell types. Finally, *unipotent cells* can serve as a continuous source of one cell type.

8.2 Embryonic Versus Adult Stem Cells

The previous paragraph described stem cells in terms of their potency. Another way to classify stem cells is by their developmental stage. There are two main categories: *embryonic* and *adult stem cells*. Embryonic cells, commonly abbreviated ESCs, are pluripotent cells that can be obtained from the inner mass of a blastocyst. They are capable of forming all cell types of the developing embryo except the placenta; so ESCs are *pluripotent* as per the definition given in the above section.

Adult stem cells, also sometimes called *resident stem cells*, reside within each type of tissue and can be either *multipotent* or *oligopotent*. They exist to replenish differentiated cells within each tissue type. Differentiated cells are very specialized units of each tissue, which constitute most of each tissue's cellular content. They are there to

◻ **Fig. 8.1** A cartoon showing stem cells yielding different specialized cells

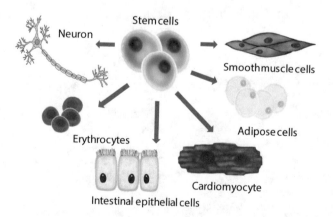

perform tissue-specific functions. Hepatocytes, for example, specialize in protein synthesis, neurons in conducting electrical impulses, smooth muscle cells in creating vessel wall tension, ventricular myocytes in heart contraction, and so on.

Fully differentiated cells are not capable of many divisions. Therefore, to replenish cells that are lost or damaged, all tissues contain a small amount of resident stem cells. Resident stem cells are responsible for maintaining a constant supply of newly differentiated cells in each tissue type. The rate of such cell renewal dramatically differs between the tissues—being very high, for example, for hematopoietic stem cells found in bone marrow, while being negligible for neurons or cardiac muscle. There are also *tissue-specific progenitor cells*. These cells give rise to specialized cells, but their capacity for self-renewal is limited.

8.3 Stem Cells in Tissue Engineering

The physiological rate of tissue renewal using progenitors or resident stem cells is not nearly sufficient to deal with large-scale tissue damage or anatomical defects. To create a sizable chunk of new tissue, the amount of newly differentiated cells needs to increase dramatically. For example, to repair an average human infarcted tissue, more than a billion new cardiac muscle cells are required.

Thus, to create an in vitro engineered tissue, one needs to first isolate stem cells and then amplify them. Among those initial sources can be embryonic stem cells (ESCs), mesenchymal stem cells (MSCs), induced pluripotent stem cells (more below), or tissue-specific resident stem cells. Stemness of these cells ensures their ability to unlimited proliferation (◻ Fig. 8.2). Once the desired amount of stem cells is achieved, they can be then differentiated into cells of choice for that specific tissue type. The last step is to seed these cells into scaffolds and culture them, so they can form a tissue (◻ Fig. 8.3).

◻ **Fig. 8.2** Colonies of undifferentiated human embryonic stem cells that can be split and passaged multiple times without losing their stemness. Scale bar: 100 micron

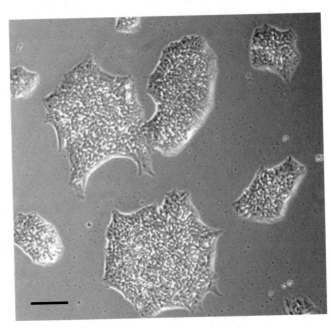

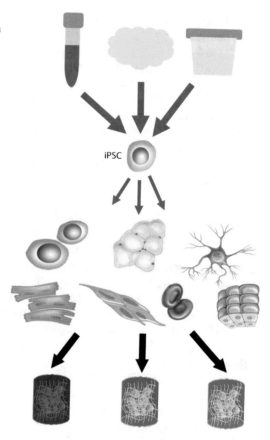

Fig. 8.3 Schematic representation of how patient's specific engineered tissues can be derived from stem cells found in blood, fat, or urine

iPSC

It is important to note that in vivo, stem cells are located within spatially highly organized stem cell niches. This enables to keep their proliferation rates physiologically controlled. Such niches are yet to be created in ex vivo engineered tissues. Therefore, in cases when a piece of newly engineered tissue contains stem cells that are not fully differentiated, its implantation poses a high risk of teratoma formation. This is because, just like in case of cancer cells, stem cells lack mechanisms of self-limiting their proliferation.

8.4 Mesenchymal Stem Cells (MSCs)

MSCs are isolated from a fully developed organism; therefore, they belong to the category of adult stem cells. All organs contain MSCs and use them for local regeneration. One of the main ethical advantages of MSCs is that their isolation does not require the destruction of the embryo. MSCs are lineage-specific and less teratogenic than ESCs, but they also have limited longevity in culture and are capable of generating only a limited number of organ-specific cell types. Current tissue engineering protocols focus mainly on adipose tissue-derived MSCs (Ad-MSCs) due to their abundance, ease of access, and relative plasticity (■ Fig. 8.4).

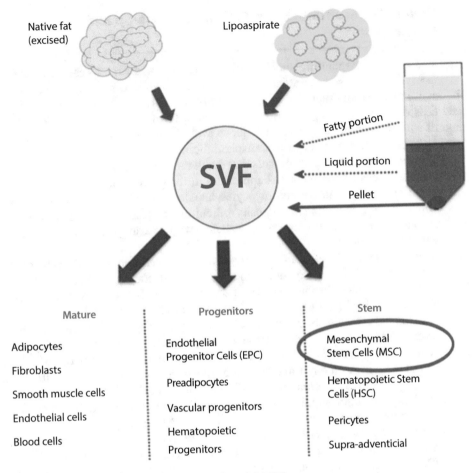

Fig. 8.4 Types of cells that can be derived from Ad-MSC

8.5 Basic Protocols for Maintenance and Growing Stem Cell Colonies

Undifferentiated stem cell colonies are usually grown using mouse embryonic fibroblast-conditioned medium (MEF-CM). More recently, several artificial media that do not use animal-derived products have also been developed. Stem cell colonies are often cultured on *layers of feeder cells*. The latter are fibroblast-like cells that provide attachment and other factors needed for stem cell maintenance. There are also protocols that enable culturing stem cells in *feeder-free conditions*. For example, primary adipose tissue-derived stem cells (Ad-MSCs) can be cultured under feeder-free conditions, using tissue culture-treated plates and daily supplementation with growth medium and human serum. Isolated, sparsely plated cells are grown to form individual colonies. Afterward, they are re-plated every 3–5 days to maintain pluripotency.

8.6 How to Differentiate Stem Cells into Tissue-Specific Phenotypes

As noted above, transplanting undifferentiated pluripotent stem cells can be dangerous because these cells are teratogenic. It was also shown that placing stem cells in the environment of an organ does not necessarily "educate" them to become specialized cells of that tissue/organ. Instead, these cells form a cluster, which does not engraft with the rest of the tissue. In order to avoid such problems, pluripotent cells need to be pre-differentiated before transplantation.

There are many published protocols that detail the sequence of steps for exposure of pluripotent cells to specific active molecules in the growth media that direct the conversion of stem cells into cardiomyocytes, hepatocytes, or neurons, to mention a few. Many factors, including the density of culturing, passaging, and re-seeding times, can affect the differentiation process. Below we give examples of three detailed differentiation protocols.

I.   *To induce the development of cardiomyocytes.* The cell line of undifferentiated human ESC (hESC) is expanded under standard feeder-free conditions using mouse embryonic fibroblast-conditioned medium (MEF-CM) supplemented with human bFGF (4 µg/L). When undifferentiated hESC colonies occupy approximately two-thirds of the surface area, they are dispersed using 0.2 mL/cm^2 Versene and re-plated to form a dense monolayer. After a confluent monolayer is formed, hESCs are switched to the RPMI-B27 medium and serially pulsed with 100 µg/L activin A on day 1 and bone morphogenetic protein-4 (BMP4, 10 µg/L) from days 1 to 5. Thereafter, the differentiating cultures are grown in the RPMI-B27 medium supplemented with insulin and vitamin A. Spontaneous beating activity commenced on days 10–20. *More details in ref* [1].

II. *To induce the development of hESC into definitive endoderm (DE) and subsequently hepatocytes.* Stem cells are cultured in serum-free conditions with the RPMI medium containing activin A (100 ng/mL), Wnt3a (50 ng/mL), 2 mM L-glutamine, and 1% penicillin/streptomycin for 24 h. Then the medium is changed to the RPMI medium with activin A (100 ng/mL), sodium butyrate (0.5 mM), and B27 supplement for 6 days. DE cells are then passaged on collagen type I-coated 12-well plates in IMDM supplemented with 20% fetal bovine serum, FGF4 (20 ng/mL), hepatocyte growth factor (HGF) (20 ng/mL), BMP2 (10 ng/mL), BMP4 (10 ng/mL), 0.3 mM L-thioglycerol, 0.5% dimethyl sulfoxide, 100 nM dexamethasone, and 0.126 U/mL human insulin for 2 weeks. *More details in ref* [2].

III. *Neuronal differentiation* can be induced via two methods: embryonic body (EB) formation or growth of adherent neurons. To make EBs, stem cells are trypsinized, counted, and grown in suspension (5×10^6 cells in 10 mL) in a 100 mm Petri dish. This culture is then placed on a shaker at 50 rpm in a 37 °C incubator. After 2 days, 20 µl of 10 mM all-trans retinoic acid is added and incubated with shaking for an additional 3 days. At day 5 formed EBs contain pre-committed neurons, which can be grown out by plating them onto poly-L-ornithine and fibronectin coated-dishes and grown in a neurobasal medium with B27, bFGF, and EGF. *Details in ref* [3].

Alternatively, to obtain adherent differentiated neurons, 10,000 ESCs are plated to an 8-well chamber slide coated with 0.1% gelatin. These cells are then cultured in Neurobasal/F12/DMEM supplemented with B27, glutamate, and N2 supplement. *Details in ref* [4].

8.7 Basic Immunology of Graft Rejection

Let's say we were able to isolate stem cells, amplify them to a desired quantity, differentiate them into a chosen cell phenotype, and finally make a piece of functional tissue (more details as to how this last step is achieved can be found in ▶ Chap. 9). When this engineered piece of tissue will be implanted into an experimental animal or a human patient, it is important to consider how the immune system of the recipient will react to it. Below we will briefly describe the basics of immunology as it applies to grafting [5]. There are three phases of acquired immune rejection. Hyperacute rejection occurs within minutes of transplantation. In this case, rejection is induced by pre-existing host antibodies that bind to the graft's antigens. This binding activates the complement system and a sequence of events ensues, including an influx of peripheral blood mononuclear cells, the formation of platelet thrombi, small vessel thrombosis, and finally, damage and/or destruction of the graft. Acute rejection occurs within 1–2 weeks after transplantation and is characterized by capillary rupture and severe graft infiltration by monocytes/macrophages, lymphocytes, and dendritic cells. This reaction is mediated by cytotoxic T lymphocytes. Chronic rejection occurs months or years after transplantation and is associated with deposits of immunoglobulin and C3 complement molecules on the basement membrane of graft cells. When using stem cell-derived grafts, all three phases of rejection should be considered, but acute rejection is the most important. Therefore, let's consider it in more detail.

Transplantation of a graft made of cells from a genetically unrelated organism (called allogeneic cells) results in a robust immune response and, consequently, graft rejection. T-lymphocytes recognize foreign antigens in the form of peptides, which are presented in association with self-major histocompatibility complex (MHC) molecules. There are two primary classes of MHC molecules. MHC class I antigens are found on the surface of every nucleated cell; they display fragments of proteins synthesized within the cell to cytotoxic T lymphocytes (CTL) or CD8+ T cells. Cell surface expression of MHC I antigens is the predominant reason for immune detection and rejection of allogeneic grafts. In comparison, MHC class II antigens are expressed only on a few specialized cell types, including macrophages, dendritic cells, and B lymphocytes. These cells are called professional antigen-presenting cells. The function of MHC II molecules is to display peptides of exogenous proteins to T-helper cells or CD4+ T cells. Direct allorecognition arises when both types of host T-lymphocytes (CD8+ and CD4+) are stimulated by donor antigen-presenting cells. This stimulation occurs through a direct interaction between host T-lymphocyte receptors and MHC I and II antigens expressed on the surface of donor cells (◻ Fig. 8.5). Indirect allorecognition occurs when donor MHC peptides are processed and then presented by host MHC II molecules. Host antigen-presenting cells internalize and break down donor MHC molecules, display those peptides onto their MHC II molecules, and present them to host CD4+ T cells. This process triggers an immune response against grafts expressing

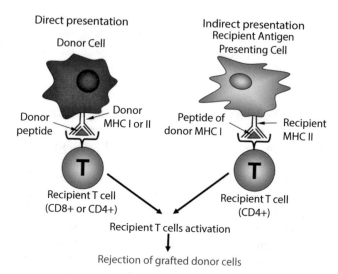

Fig. 8.5 Schematic representation of direct and indirect recognition of donor antigens to the recipient T cells

donor MHC, both I and II. All-in-all, the presence of MHC molecules is the defining factor that largely determines the degree of acute graft rejection.

8.8 Induced Pluripotent Stem Cells

Until very recently, our ability to amplify and differentiate human stem cells was focused on stem cells obtained from discarded human embryos, freshly excised pieces of tissue from surgical procedures, or from recently deceased donors. In all these three cases, engineered tissues made from these cells will be rejected by the recipient's immune system as per the previous paragraph. The only solution to this problem was to (a) find the best match as far as immunocompatibility and (b) subject graft recipient to a lifelong regimen of immunosuppression drugs. Fortunately, sustained research efforts in the cell biology field led to a giant breakthrough. In 2006, a team of Japanese researchers led by Shinya Yamanaka published a study [6] in which they successfully reprogrammed intact mature cells from connective tissue into stem cells by introducing only a few genes. Synthesis and action of four regulatory proteins (OCT3/4, Sox2, c-Myc, and Kfl4) were sufficient to govern such de-differentiation. These regulatory proteins have since been called Yamanaka factors. Next came a possibility of what is called "direct reprogramming," a procedure that enables one to convert specialized cells of one type into cells of another tissue without going back to stem state. This approach requires genetic or chemical modifications and is called re-differentiation. It allows to save time in generating patient-specific cells and is currently one of the most actively developing branches of cell biology [7, 8].

Stem cells that were reprogrammed from a patient's own fully differentiated cells are called *induced pluripotent stem cells* or *iPSC*. Today, multiple types of differentiated cells have been reprogrammed into iPSC, including fibroblasts, white blood cells, adipocytes, epithelial kidney cells, and many others.

Discovery of iPSC means that it is now possible to derive a person's own pluripotent stem cells from small samples of their blood, skin, fat, or even urine. iPSC obtained from these samples can be then amplified in vitro and re-differentiated into desired specific types of cells. In layman terms, it is now possible to derive a person's own nerve or heart cells from urine sample. This has been a truly exciting development from the perspectives of both science and medicine as it turns humanity's dream of creating entirely new organs from a person's own cells into a reality. It has also opened the doors to what is now called personalized drug testing. The latter uses engineered tissue made from an individual's own cells to test the effects of the drugs and their combinations.

As of today, protocols to create engineered tissues from iPSC are very time-consuming and labor-intensive. They are also very expensive. However, every year, thanks to efforts of thousands of researchers from across the world, these protocols are becoming more and more affordable, reproducible, and scalable. This progress enables the hope of using iPSC-derived tissues for clinical treatments in a not-so-distant future.

8.9 Immunohistochemistry/Immunostaining

Epitope recognition by specific antibodies developed by the body can be used as a method to identify certain proteins within or on the surface of the cells. It involves immunization of an animal with an epitope of interest (which can be a full protein or its fragment). Immunoglobulins that develop against the epitope are called *primary* antibodies. *Secondary* antibodies can then be used to recognize *primary* antibodies based on antibody type and the animal in which the primary antibody was developed in. The secondary antibody can have a fluorescent label or an enzyme attached to it, enabling colorimetric or fluorescence-based detection. ◘ Figure 8.6 shows the main players involved in a typical immunostaining protocol. To obtain good quality immunostaining, it is important to optimize fixation methods, timing and concentration of blocking agents, and concentration and duration of exposure to both primary and secondary antibodies. It is also recommended to have both positive (i.e., a sample in which epitope of interest is present in high amounts) and negative controls. For the latter, the same exact sequence of steps is applied, but the primary antibodies are omitted. This enables ruling out the non-specific staining by the secondary antibodies.

8.10 Choice of Antibodies

Primary antibodies have to be raised in a host species that is different from the species of the antigen in order to avoid cross-reactivity with endogenous immunoglobulins. For example, let's say we are trying to stain mouse cells for the presence of connexin 43. The choice should be a primary antibody that is raised in a species other than mice. For example, the primary antibody raised in goat (or rabbit or horse) would be a good choice and the secondary antibody should be then anti-goat IgG.

◘ Fig. 8.6 A cartoon showing a basic
sequence of epitope labeling used in
immunostaining protocols

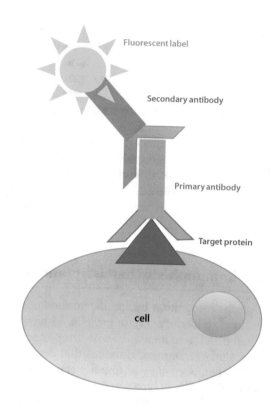

Fluorescent label

Secondary antibody

Primary antibody

Target protein

cell

Session I

Demonstration

The instructor shows steps involved in the isolation of adipose mesenchymal stem cells from adipose tissue and immunostaining for stem cell markers. Briefly, a fresh visceral or subcutaneous animal fat is washed with PBS to remove most of blood cells. Fat tissue is minced using a scalpel into slurry and transferred into a 50 mL tube with 1–2 mg/mL collagenase I solution in PBS. After 30 min incubation using vortexing, serum-containing media are added to stop the digestion, following 800 xg for 10 min. Top fatty layers are discarded, while the remaining clear solution is filtered through a 100-micron filter and centrifuged for 5 min at 200 g. Pellet is resuspended in erythrocyte lysis buffer (155 mM NH_4Cl, 10 mM $KHCO_3$, 0.1 mM EDTA) for 5 min, followed by another round of centrifugation and cell count. Cells are seeded at ~2000 cells/cm^2 and cultured using available standard cell culture media supplemented with antibiotics.

Homework

Students are tasked to search the literature for most suitable antibodies and protocols to stain stem cells from the demo session. A list of available primary and secondary antibodies is given to students to narrow down their choices.

Session II
Team Exercises

Students stain adipose explant stem cell cultures with stemness markers of their choice, followed by imaging using a fluorescent microscope.

Homework

Students are tasked with making a PowerPoint presentation that includes images of stained cell cultures with and without primary antibodies.

ⓘ Sample Protocol
Immunostaining

1. Fix samples by incubating in ice-cold 1:1 methanol-acetone solution for 10 min. Rinse twice with PBS.
2. Block specimen for 60 min in blocking solution (PBS, 5% normal serum, 0.3% Triton™ X-100). Use serum from the same species as the secondary antibody. Alternatively, use 1% albumin.
3. Aspirate blocking solution and apply diluted primary antibody. Incubate overnight at 4 °C.
4. Rinse three times in PBS for 5 min each. Incubate specimen in fluorochrome-conjugated secondary antibody diluted in PBS, 1% BSA, 0.3% Triton™ X-100 for 1–2 h at room temperature in dark. Rinse in PBS.
5. Cover samples with a small amount of antifade reagent or Mowiol. Place a coverslip on the top and seal with clear nail polish.
6. Examine samples using appropriate excitation/emission settings. Store slides at 4 °C protected from light.

> **Take-Home Message/Lessons Learned**
>
> After reading this chapter and performing the requested assignments and exercises, students should be able to:
> - Define degrees of cell stemness and terms to describe them
> - Understand three phases of immune rejection and the key players involved in rejection of implanted engineered tissue
> - Choose appropriate primary and secondary antibody for a specific sample
> - Name four Yamanaka factors
> - Understand the complexity of differentiation protocols and their cell-type specificity

Self-Check Questions

❓ Q.8.1. Choose the correct sequence of terms describing the decreasing capacity of stem cells to give rise to different phenotypes.
   A. Totipotent, pluripotent, multipotent, oligopotent, unipotent
   B. Pluripotent, totipotent, multipotent, oligopotent, unipotent
   C. Multipotent, totipotent, pluripotent, unipotent, oligopotent
   D. Pluripotent, totipotent, multipotent, oligopotent, unipotent

? Q.8.2. All the following statements about MHC class I are correct, EXCEPT
   A. MHC class I antigens are found on the surface of every nucleated cell.
   B. MHC class I display fragments of proteins synthesized within the cell to cytotoxic T lymphocytes (CTL) or CD8$^+$ T cells.
   C. The function of MHC I molecules is to display peptides of exogenous proteins to T-helper cells or CD4$^+$ T cells.
   D. Cell surface expression of MHC I antigens is the predominant reason for the rejection of allogeneic grafts.

? Q.8.3. Discovery of induced pluripotent stem cells (iPSCs) is considered a scientific breakthrough mainly because it
   A. Suggested that stem cells can give rise to different types of specialized cells
   B. Overthrew long-standing dogma that differentiation process cannot be reversed
   C. Demonstrated that stem cell differentiation can occur in both animal and human subjects
   D. Was the first time that stems cells' existence was experimentally shown

? Q.8.4. When stem cell colonies start to merge, it is critical to disperse and re-seed stem cells at lower densities because
   A. Stem cells will start to differentiate
   B. pH of the media will become more alkaline
   C. The amount of cells will exponentially increase
   D. Contamination will be imminent

? Q.8.5. Choose the correct statement.
   A. There are five phases of acquired immune rejection.
   B. Immunoglobulins that develop against the epitope of interest are called primary antibodies.
   C. Acute rejection is mediated by cytotoxic T macrophages.
   D. Transplantation of undifferentiated pluripotent stem cells is safe and can never lead to teratomas.

References and Further Reading

1. Z. Karabekian et al., HLA class I depleted hESC as a source of hypoimmunogenic cells for tissue engineering applications. Tissue Eng. Part A **21**(19–20), 2559–2571 (2015)
2. Y. Duan et al., Differentiation and characterization of metabolically functioning hepatocytes from human embryonic stem cells. Stem Cells **28**(4), 674–686 (2010)
3. C. Chatzi et al., Derivation of homogeneous GABAergic neurons from mouse embryonic stem cells. Exp. Neurol. **217**(2), 407–416 (2009)
4. Q.L. Ying, M. Stavridis, D. Griffiths, M. Li, A. Smith, Conversion of embryonic stem cells into neuroectodermal precursors in adherent monoculture. Nat. Biotechnol. **21**(2), 183–186 (2003)
5. Z. Karabekian, N.G. Posnack, N.A. Sarvazyan, Immunological barriers to stem-cell based cardiac repair. Stem Cell Rev. **7**(2), 315–325 (2011)

6. K. Takahashi, S. Yamanaka, Induction of pluripotent stem cells from mouse embryonic and adult fibroblast cultures by defined factors. Cell **126**(4), 663–676 (2006)

7. J. Kim et al., Direct reprogramming of mouse fibroblasts to neural progenitors. Proc. Natl. Acad. Sci. **108**(19), 7838–7843 (2011)

8. T. Noda et al., Direct reprogramming of spiral ganglion non-neuronal cells into neurons: Toward ameliorating sensorineural hearing loss by gene therapy. Front. Cell Dev. Biol. **6**, 16 (2018)

Scaffolds and Tissue Decellularization

Narine Sarvazyan

Contents

© Springer Nature Switzerland AG 2020
N. Sarvazyan (ed.), *Tissue Engineering*, Learning Materials in Biosciences,
https://doi.org/10.1007/978-3-030-39698-5_9

What You will Learn in This Chapter and Associated Exercises

The goal of this chapter is to give a broad overview of scaffolds including artificial and biological ones. The key questions to ask when selecting scaffold for engineered tissue as well as how to re-populate scaffold material with cells will be considered. Process of decellularization using examples of different organs will be shown. Students will then practice decellularization of an organ of their choice based on the selected online protocol.

9.1 Overview

Most of the mammalian cells are anchorage-dependent. This means that in order to survive, they need a substance for attachment. *Scaffolds* provide structural and mechanical support to cells, allowing their attachment and thus the development of 3D tissue. An ideal scaffold would be the extracellular matrix (ECM) of the target tissue itself, in its native conditions. However, because of its complexity, diverse functions, and the fact that it has a dynamic nature, it is difficult to fully recreate the ECM. Therefore, different scaffold materials and fabrication methods try to mimic, as much as possible, the functions and structure of the native ECM in which cells reside in vivo.

There is a great variety of different scaffold materials, depending on their purpose and/or target tissue to be engineered. Depending on their chemical composition, materials used in scaffolding can be classified into four broad categories: ceramics, polymers, composites, and decellularized matrix. *Ceramics'* main components are inorganic substances (metal compounds, calcium salts), and they are most often used in orthodontics. *Polymers* can include both molecules that exist biologically as well as synthetic ones. Because of a wide range of the mechanical properties of scaffolds that they can create, polymer-based materials are being used to re-create almost all kinds of tissues. *Composite materials* represent a mix of ceramics and polymers and are used both in orthopedics and orthodontic tissue engineering. Finally, there is a *decellularized matrix* that contains a natural mix of native tissue polymers and, in case of decellularized bone, can also include inorganic components.

Biomaterials can also be classified by their origins—natural or synthetic. Natural biomaterials (such as collagen, chitosan, alginate, etc.) have the advantage of biocompatibility and biodegradability. Synthetic biomaterials can be both non-degradable and degradable. Their advantages include the possibility of chemical modifications that enable to control the rates of their degradation, release of bioactive components, or cell attachment properties.

The process of the creation of scaffolds starts with the choice of proper biomaterials, which is then followed by scaffold *fabrication*. During this process, they undergo physical and/or chemical modifications to meet specific requirements of the target tissue and become usable for tissue engineering purposes. After creating a scaffold with necessary properties, it can either be populated with cells and cultured in vitro until the desired tissue is developed, or it can be grafted to the patient without the preliminary addition of cells so that the host cells can infiltrate and grow in the scaffold. In this chapter, we will talk mainly about the initial materials from which scaffolds are made, while ▶ Chap. 10 will focus on how these materials can be given a specific shape or mechanical properties.

9.2 Key Questions to Ask When Choosing Scaffold Material

Any material to be used as a 3D scaffold must satisfy the following three requirements. First, it should be biocompatible and should be able to support the influx of nutrients and the efflux of cellular waste. The second requirement is that the scaffold material must support cell attachment. This is pivotal since cell seeding typically is performed under the flow of cell-containing solution through the material. To enhance the attachment and before recellularization, scaffold material may be coated with collagen I, collagen IV, fibronectin, laminin, or other adhesion proteins. Finally, the third key aspect of the scaffold material is its stiffness. This feature is important since the material has to endure the mechanical stress when placed into bioreactors or inside the body but should be soft and/or porous enough for cells to penetrate into it. The key questions to be asked when choosing scaffold material for creating 3D tissue include the following:

1. Will the chosen type of cells attach to selected scaffold material?
2. Is it hydrophilic?
3. What is scaffold porosity?
4. Will it be degraded upon implantation?
5. Are its degradation products toxic?
6. Can it be made into injectable material?
7. What are the scaffold's mechanical properties?
8. Can/when it be vascularized?
9. Is it immunogenic?
10. Will it melt or solidify at body temperature?

9.3 Means to Seed Cells into Scaffold Material

After cells are amplified to a required amount and scaffold material is ready, the next step is to introduce cells into the 3D scaffold structure. This can be done by several means. Specifically, cells can be:

1. Directly injected into scaffold material
2. Mixed into scaffold material before it polymerizes
3. Seeded-in by sedimentation, be it passively or by using centrifugal force
4. Distributed by placing scaffold in a cell-containing media while it is being constantly stirred
5. Infused into a scaffold by perfusion with cell-containing media being passed through scaffold material
6. Introduced via natural vessels (in case of decellularized scaffolds) or synthetic vasculature-like channels
7. 3D printed, layer-by-layer while mixed with a scaffold material

9.4 Hydrogels

Hydrogels represent 3D polymeric networks filled with water (up to 90–99%, depending on the concentration of the polymer). Thus, they are hydrophilic and are able to safely incorporate proteins, cells, growth factors, and other biological entities. Because of their viscoelastic properties, they can be directly injected into the lesion site of the patient. Hydrogels can be made of different materials: synthetic

(polyethylene glycol, polyacrylamide, polydimethylsiloxane, etc.) or natural (collagen, gelatin, alginate, chitosan, hyaluronic acid, etc.). They further undergo gelation
process, which can be carried out by covalent (e.g., UV curing) or physical (e.g.,
changing the temperature or pH) cross-linking. Covalent cross-linking creates additional bonds between polymer chains. The fate of the cells (their attachment, proliferation, differentiation) is highly dependent on different characteristics of the
hydrogel scaffolds, including hydrogel stiffness and elasticity.

As mentioned above, hydrogels can be composed of multiple different polymers.
They can also be impregnated with cell-adhesion ligands (e.g., fibronectin, laminin)
that support cell attachment, spreading, and proliferation. Alternatively, or in addition, they can be enriched with signaling molecules (e.g., bone morphogenic proteins
or transforming growth factor-beta 1), which can guide cell differentiation to the
desired phenotype within the hydrogel scaffold. Having all these multiple variables
makes it increasingly hard to track cell behavior within scaffolds made from different
materials. Recently, an interesting approach to tackle this issue was suggested by Rui
Reis lab (Minho, Portugal). It enables the creation of hydrogel-based or other types
of polymer fibers with their components mixed in a controlled, gradient type of fashion. These fibers are then rolled into a cassette-like pattern. This way multiple conditions can be analyzed at once. For example, one end of the fiber can be 100% agarose
and the other 100% alginate gel. In between, all possible ratios of these two hydrogels
will be created. Assessment of cell proliferation or differentiation status inside such
fiber can be then done all at once using staining with either functional or viability
probes. If required, more than two components can be mixed together so the whole
range of variables can be analyzed during a single experiment (◘ Fig. 9.1).

Similar to other types of scaffolds used in tissue engineering, degradability plays
an important role in case of hydrogel-based scaffolds. Degradation is required in
order to allow the cells within the scaffold to replace it with ECM. Degradation can

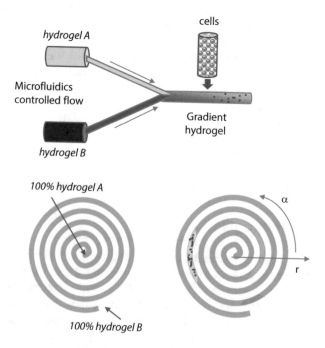

◘ **Fig. 9.1** An example of a microfluidics-based approach to screen different compositions of hydrogel material affecting cell behavior. After the extruded fiber is rolled into a cassette, markers for different features (shown as green, orange, or yellow colors) can be applied and imaged at once

be provided by different approaches, including hydrolysis or enzymatic degradation, which incorporates enzymatically cleavable sites (e.g., matrix metalloproteinase cleavage site) into the hydrogels. It is important to note that degradation leads to changes in the viscoelastic properties of the scaffolds, which is a key factor that influences the fate of the cells embedded within scaffold material.

9.5 Natural Scaffold: Decellularized Tissue (DCT)

Cells reside inside a dense network of protein fibers, proteoglycans, and glycosaminoglycans, collectively called extracellular matrix (ECM), as discussed in ▶ Chap. 3. ECM provides both structural support and signaling molecules that influence cell behavior. One way to obtain a natural tissue scaffold is to remove cells leaving behind just ECM (◘ Fig. 9.2). In reality, complete removal of cells is not possible without disturbing ECM architecture. Therefore, a certain balance between gentle and aggressive treatment has to be achieved. There are multiple uses for DCT scaffolds, specifically

1. Implanting DCT for structural repair (valve replacement, organ wrapping, vessel repair)
2. Implanting DCT with the goal of reseeding it with patient's own cells (skin grafts, vessel repair)
3. Using it as a non-immunogenic scaffold to be seeded with patients' own cells to grow small pieces of engineered tissue for personalized pharmacological testing

◘ **Fig. 9.2** The appearance of rat lungs throughout 24 h DCT protocol: **a** heart and lung excision, **b** tracheal cannulation, **c** partial DCT, **d** complete DCT

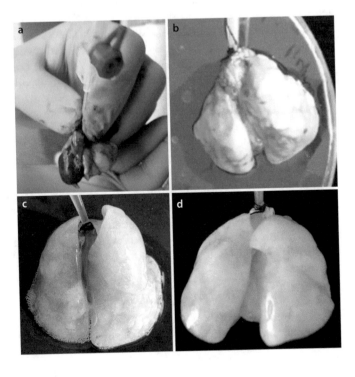

4. As a raw material that can be homogenized to be used as an injectable hydrogel or bioink
5. Using for research purposes such as, for example, reseeding it with different cell types to examine how they will be homed to their natural anatomical niches

The use of DCT was first reported in 1973 as a technique to preserve skin layers intended to be used as a protective barrier for burn patients [1]. Today there are numerous approved clinical products based on DCT. Interested readers can search online for the detailed description of DCT and ECM derived from them [2]. They can be easily stored and preserved and, thus, are considered "off the shelf" products.

Another interesting approach is the use of DCT from plant tissues. Decellularized leaves or other plant materials have been shown to support the growth of mammalian cells, especially if coated with adhesion proteins. In fact, a commercial product called GrowDex was recently developed from a birch tree. It is being promoted as a ready-to-use hydrogel with tunable viscosity that requires no cross-linking, no gelation, no sonication, or any other steps for it to gel. It can be mixed with cells and used for 3D cultures and organ-on-a-chip models, in drug release studies, or as a bioink in 3D printing.

9.6 Ways to Obtain DCT

Ideally, one wants to remove all the cells while leaving ECM architecture intact to a maximal degree. This is because if the ECM macrostructure is disrupted in a non-natural way, it can elicit immune or foreign body response and negatively impact the ability to re-seed DCT with cells. Therefore, it is important to achieve a balance between quick and successful decellularization (i.e., the high concentration of detergents, enzymes, or salts) and keep ECM relatively intact. For organs that can be cannulated, such as heart, liver, or kidney, the best results can be achieved via organ perfusion, since it gives full access to capillary bed allowing lower concentrations of detergents, enzymes, or salts. For organs such as skin, bone, tendons, fat, or other tissues where cannulation is not possible, samples are usually cut into smaller or thinner pieces followed by mechanical agitation in the presence of enzymes, salts, or detergents. Mechanical forces such as pressure or electroporation are also sometimes used.

9.7 Reagents to Decellularize

Several common treatments are used to remove cells from different types of tissue. After these steps, DCT usually undergoes multiple cycles of washing using double distilled water (ddH$_2$O) to get rid of the remaining agents. Even a small concentration of such agents can be cytotoxic and negate the beneficial clinical properties of DCT. Below we consider the main steps and/or treatments involved in producing DCT.

Freezing   Below −4 °C produces ice crystals inside the cells. These crystals break the cell membrane, and so upon returning to room temperature, the first step of disrupting cell structure is uniformly accomplished. Importantly, many freezing and defrosting steps can also affect the integrity of the ECM structure, so these procedures have to be applied in moderation. Notably, heating tissue (which also kills cells) is not used in DCT protocols since it leads to denaturation of ECM proteins.

Hypo and hyperosmotic solutions Treatment of tissue with alternating hypo- (water) or hypertonic (>0.5 M NaCl) solutions leads to cell lysis by the osmotic shock, which disrupts DNA-protein interactions. However, this treatment does not effectively remove cellular residues. The main benefit includes minimal changes in matrix molecules and architecture. Usually, tissues are immersed alternately in hyper- and hypotonic solutions through several cycles. Similar steps are also effective in removing detergents or other decellularization agents.

Solvents *Alcohol, methanol, chloroform,* and *acetone* cause cell lysis by dehydration. They solubilize and remove lipids. They are effective in removing cells from dense tissues but also crosslink and precipitate proteins, including collagen. They also make scaffolds stiffer by physically working as a fixative.

Acids and bases Some protocols use acids (such as acetic or peracetic acid). They effectively solubilize cytoplasmic components of cells but can also disrupt nucleic acids and denature proteins. Bases (e.g., calcium hydroxide, sodium hydroxide) are too harsh for most tissues due to their destruction of collagen fibers but can be used to remove hair from dermis samples during the early stages of decellularization.

Chelators (EDTA, EGTA) Chelating agents bind calcium loosening cell-to-cell connections. They are not effective when used alone but can be useful in conjunction with other reagents.

Enzymes *Nucleases* can help to chop and remove nucleic acids after cell membranes are disrupted. *Proteases* (trypsin, dispase) are effective in cell detachment, but at longer exposures they also destroy ECM structure.

Detergents Include non-ionic, ionic, or zwitterionic agents. Non-ionic detergents, such as Triton X-100 or TWEEN-20, disrupt DNA-protein, lipid-lipid, and lipid-protein interactions while maintaining native protein structures. Ionic detergents, such as sodium dodecyl sulfate (SDS) or sodium deoxycholate, completely solubilize cell and nucleic membranes and can fully denature proteins. Zwitterionic detergents, for example, 3-[(3-cholamidopropyl) dimethylammonio]-1-propanesulfonate (CHAPS), have a net-zero electrical charge on the hydrophilic head groups, which protects the native state of proteins during decellularization and exhibits properties of both ionic and non-ionic detergents.

9.8 Checking Vascular Integrity

When an organ is cannulated to be perfused with decellularization agents, it is important to visually confirm that perfusion cannula entered a major vessel, and that perfusion flow is evenly distributed. For this, one can use a diluted solution of food

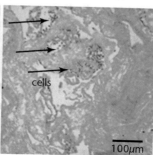

| Decellularized liver | Vascular integrity check | Histology of re-seeded DCT |

▣ Fig. 9.3 After decellularization, the tissue becomes transparent as shown here using rat liver as an example. A food dye can be added to the perfusion cannula to delineate the vessels. Scaffolds can be then reseeded with cells and analyzed histologically

coloring or other non-toxic dye that can outline the vascular bed before and after the decellularization process (▣ Fig. 9.3). To remove such staining, tissue can be subjected to another round of perfusion using low concentration detergents such as SDS or Triton.

9.9 Checking the Quality of Decellularized Tissue

After DCT is completed, it is important to document three main outcomes. This includes testing: (i) the degree to which cellular content has been removed, (ii) whether DCT is free from remnants of detergents and salts, and (iii) the level of integrity of the remaining ECM. The former outcome is important since any remaining cells or their pieces can elicit a potential immunogenic response and have to be eliminated as much as possible [3]. Usually, highly sensitive assays for nucleic acids are used to quantify the presence of any remaining nuclear material, examples being Picogreen or ethidium bromide staining. Count of remaining cell nuclei in histological samples is also sometimes used to illustrate cell removal; however, the latter approach is much less quantitative. The second outcome—documenting the removal of detergents—is also important as it will interfere with DCT recellularization. To verify the integrity of the ECM scaffold, it is best to image a 3D microstructure of DCT by using scanning electron microscopy. A good decellularization protocol yields a regular ECM structure with visible pores (lacunae) where cells resided before. Studies have shown that pore sizes of 30–40 microns are the best for reseeding of DCT or artificial scaffolds. Scaffold proteins can also be stained with different histological dyes, in which case specific stains can point to the abundance of individual ECM components, that is, elastin, collagen, or glycosaminoglycans. The final steps of successful DCT protocol should include checking in vivo immunocompatibility (i.e., lack of rejection upon implantation) and cell compatibility. The latter can be done in vitro and includes assessment of whether cells can survive, proliferate, differentiate as well as perform their specific function within DCT (i.e., contract or produce tissue-specific proteins).

Session I

Demonstration

The instructor performs liver cannulation via portal vein followed by 1% SDS perfusion. The latter visibly clears the tissue from cellular content. Afterward, organ is perfused with food dye to show major vessels. Stored cannulated organs from week 2 can also be defrosted and used for the demo or as a backup.

Homework

Teams are tasked with searching the literature to find the simplest and most suitable protocols to decellularize their organ of choice and how to record the endpoints of the decellularization process.

Session II

Team Exercises

Each team chooses an organ that can be perfused and an organ that can be immersed/shaken. DCT protocol is then chosen based on literature search and available reagents. Tools to cannulate the first organ and dissect the other need to be prepared. Solutions for all the steps need to be made, and team members need to discuss who will be coming and when to change the media and take pictures for a short presentation. Things need to be done quickly and on ice for effective cannulation to avoid blood clotting. Choice of reagents can include, for example, distilled water, trypsin, EGTA, liquid nitrogen, NaCl, Triton X-100, or SDS. Once cannulated, organs can be perfused via perfusion pump or gravity fed perfusion. In either case, it is useful to have a larger bucket to collect effusate, or to place the container in the sink for fluid not to spill over onto the floor. Students need to check for evidence of fluid coming out of the organ and adjust either pump flow rate or height of the vessel with fluid in case of a gravity-based system accordingly.

Homework

Team members take turns during the next few days to finish the decellularization process and to record its endpoints. PowerPoint slides showing tissue micro and macro appearance have to be prepared.

ⓘ Sample Protocols

Decellularization protocol for rat heart (modified from [4])

1. Run antibiotic-antimycotic containing PBS (AA-PBS) through peristaltic pump tubing, followed by 10 min of 70% ethanol and another round of AA-PBS.
2. Heparinize the animal and wait for 15–20 min.
3. Anesthetize the animal as per the approved animal protocol. For adult rats, this can be done with ketamine and xylazine intraperitoneal injection or inhalant-based anesthesia. Steps 2 and 3 should be performed with the help of the instructor.
4. Spray and wipe the abdomen of the animal using 70% EtOH.
5. Perform the median sternotomy to open the pericardium and remove the heart from the chest. Insert aortic cannula into the ascending aorta for retrograde coronary perfusion.

6. Perfuse heart with cold PBS-AA solution to clear from the blood.
7. Start perfusion with 0.5% SDS in deionized water.
8. Periodically image heart using a color camera to record the clearing of cellular content. Record the timing of each image.
9. After approximately 20–24 h of 0.5% SDS perfusion, the heart should become visibly clear. If not, continue perfusing with SDS.
10. Switch tubing to deionized water for 30 min.
11. Perfuse with 1% Triton X-100 in deionized water for 30 min of perfusion.
12. Perfuse with AA-PBS for an additional 2–3 h.
13. Sterilize scaffold by 10–15 min under available UV sources.
14. Store in AA-PBS at 4 °C to be used for histology or recellularization experiments.

If a rat is not available, the fish heart can be decellularized in a similar way (◻ Fig. 9.4). The only difference is the choice of the cannulation site, which is different between the two animals (as per ▸ Chap. 2).

Decellularization protocol for small intestine (modified from [5])
1. Perform steps 1–4 as above.
2. Extend incision on the abdomen to fully expose the intestinal area. Wet intestinal tissues with additional PBS to avoid dehydration.
3. Locate the superior mesenteric artery (SMA). It forms about a 90° angle from the aorta and leads toward the intestine. Use a 27G cannula or similarly sized custom-made cannula to enter the SMA from the aorta. Advance the plastic tube of the cannula into the SMA and secure using sutures.
4. Confirm cannulation by injecting a small amount of AA-PBS and release cannula by incising aorta proximal and distal to SMA.
5. Free the intestine from any connective tissue connections, cut it at the jejuno-duodenal junction, and transfer Petri dish with PBS/AA.
6. Insert plastic tubing to the proximal end of the intestine, connect it to a large syringe, and flush out its inner content.
7. Perfuse with deionized water at slow speed (1–2 mL/h). Make sure there are no bubbles. Continue for 24 h in a cold room or fridge.
8. Switch to 4% sodium deoxycholate at room temperature for 3–4 h, then AA-PBS/AA for 30 min.

◻ **Fig. 9.4** The appearance of sea bass heart before and after 24 h decellularization with 0.5% SDS

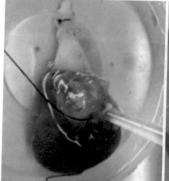
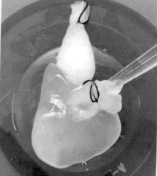

9. Perfuse with DNase-I for 2–3 h, followed by AA-PBS for an additional 2–3 h.
10. Sterilize scaffold for 10–15 min under available UV sources.
11. Store scaffold in AA-PBS at 4 °C to be used for histology or recellularization experiments.

Take-Home Message/Lessons Learned

After reading this chapter and performing the requested assignments and exercises, students should be able to:

- Name the basic types of scaffold materials used in tissue engineering
- Ask questions to select the most suitable scaffold material
- Based on the given information, choose detergents, salts, or other treatment to decellularize an organ or a tissue of choice
- Recognize the most common methods to verify ECM integrity and absence of cellular material within decellularized tissue
- Perform simple decellularization protocol

Self-Check Questions

? Q.9.1. The role of DNAse during decellularization process is to
   A. Cleave nucleic acids
   B. Dissolve lipids
   C. Collapse ECM structure
   D. Hydrolyze peptide bonds

? Q.9.2. Each of the detergents listed on the left can be assigned to a subclass listed on the right, EXCEPT
   A. Triton X-100   Non-ionic
   B. SDS   Ionic
   C. CHAPS   Zwitterionic
   D. TWEEN-20   Anionic

? Q.9.3. For most cell types, reseeding cells into DCT-based or artificial scaffolds works the best when the pore size of the scaffold is about
   A. 1–2 microns
   B. 10–20 microns
   C. 30–40 microns
   D. 100–200 microns

? Q.9.4. Choose *the least important outcome* to document after completion of tissue decellularization.
   A. The degree to which cellular content has been removed
   B. Whether DCT is free from remnants of detergents and salts
   C. Whether it retained the same degree of stiffness as original tissue
   D. The level of integrity of the remaining ECM

❓ Q.9.5. The following treatments are commonly used during DCT steps, EXCEPT

A. Cooling below 4°C
B. Heating above 50°C
C. Increasing osmolarity
D. Decreasing osmolarity

References and Further Reading

1. R.A. Elliott Jr., J.G. Hoehn, Use of commercial porcine skin for wound dressings. Plast. Reconstr. Surg. **52**(4), 401–405 (1973)
2. L.T. Saldin, M.C. Cramer, S.S. Velankar, L.J. White, S.F. Badylak, Extracellular matrix hydrogels from decellularized tissues: Structure and function. Acta Biomater. **49**, 1–15 (2017)
3. L.J. White et al., The impact of detergents on the tissue decellularization process: A ToF-SIMS study. Acta Biomater. **50**, 207–219 (2017)
4. H.C. Ott et al., Perfusion-decellularized matrix: Using nature's platform to engineer a bioartificial heart. Nat. Med. **14**(2), 213–221 (2008)
5. P. Maghsoudlou, G. Totonelli, S.P. Loukogeorgakis, S. Eaton, P. De Coppi, A decellularization methodology for the production of a natural acellular intestinal matrix. J. Vis. Exp. **80** (2013). https://doi.org/10.3791/50658

Casting and 3D Printing

Vahan Grigoryan and Narine Sarvazyan

Contents

© Springer Nature Switzerland AG 2020
N. Sarvazyan (ed.), *Tissue Engineering*, Learning Materials in Biosciences,
https://doi.org/10.1007/978-3-030-39698-5_10

What You will Learn in This Chapter and Associated Exercises

Students will learn about the most common approaches to create 3D shapes from biologically compatible scaffold materials including 3D printing. Advantages and shortcomings of different bioinks and use of support bath will also be discussed. Students will then practice making molds and casts using scaffold material of their choice.

The ultimate goal of the TERM field is to implant engineered tissue made from cells and scaffold material to cure a patient (◻ Fig. 10.1). Tissue engineering is a rapidly evolving field, and we encourage students to read most up-to-date articles on how to create and combine scaffold material with cells. Several excellent reviews on this subject are available via free online access [1–3]. This chapter outlines just the most general and well-known approaches.

10.1 Multilayer Cell Sheets

This approach relies on the properties of thermally regulated polymeric materials, such as poly(N-isopropylacrylamide). At temperatures above 32 °C (such as that in the human body or in a cell culture incubator), poly(N-isopropylacrylamide) is cell-adhesive; so when cells are seeded on it, they adhere to the surface, grow normally until full confluency, and secrete the ECM themselves. However, when the temperature falls under 25 °C, the substance loses its adhesive features, and thus the cell sheets can be lifted without the need for trypsinization. Multiple cell sheets can then be combined with each other and used for transplantation (◻ Fig. 10.2). Cells can also be seeded on other types of support material including collagen, silk, or cellulose sheets and then be stacked together to form thicker tissues.

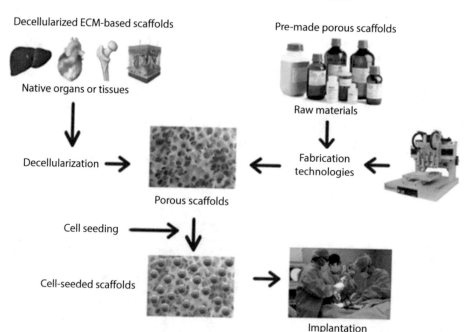

◻ **Fig. 10.1**   Schematics of using decellularized matrix or artificial materials to make and implant premade porous scaffolds

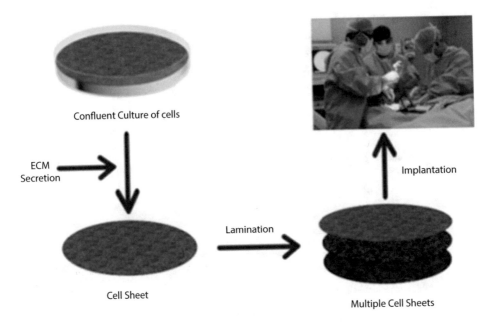

Fig. 10.2 Schematics of cell-sheet-based scaffolding

10.2 Casting

Before polymerization, polymer-based scaffolds can be fluid or pliable so they can be cast to conform to a desired three-dimensional shape (**Fig. 10.3**). Afterward, they can be stiffened chemically by altering pH (an example being collagen), increasing calcium concentration (an example being alginate), or UV crosslinking (an example being GelMA—more on this discussed below). In the case of thermoplastic materials, a polymer can be heated above its glass transition or melting point to become liquid and then poured into the mold. When cooled, the sample solidifies taking the shape of the mold. When it is possible to keep polymer temperature within the physiological range, then cells can be included in the material to be cast. If not, cells can be seeded on top of the cast polymer or introduced by other means (see ▶ Sect. 9.3).

10.3 Creating Porous Scaffolds

Fabrication techniques to create porous scaffold include melt-molding, freeze-drying, phase separation, electrospinning, self-assembly, and textile technologies. These pores or simply spaces between fibers are essential for cells to colonize these scaffolds and convert them to pieces of engineered tissue.

Melt-molding is a classical method of creating pores inside cast material. Porosity is achieved by mixing the substance with *porogens*—masses of particles that create pores in the scaffolds. Salt, sugar, gelatin, or paraffin can serve as porogens. In some cases, two different immiscible thermoplastic polymers can be combined with each other, with one serving as a material for the future scaffold and the other as a porogen. There are also other approaches for creating porosity, such as the gas foaming

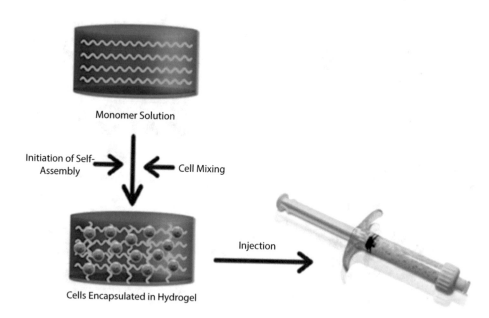

Monomer Solution

Initiation of Self-Assembly → ← Cell Mixing

Injection →

Cells Encapsulated in Hydrogel

Fig. 10.3 Schematics of hydrogel-based scaffolds

technique, which uses foaming agents (e.g., sodium bicarbonate, CO_2) to create gas bubbles for providing porosity in the scaffolds.

As an example of melt-molding technique can refer to the preparation of scaffolds from poly(lactic-co-glycolic acid) (PLGA), with gelatin microspheres of the desired size used as a porogen. This mixture is poured into a Teflon mold and heated above the polymer's glass transition temperature. Afterward, the PLGA-gelatin composite is placed in distilled-deionized H_2O. Since gelatin is water-soluble, it can be leached out, leaving behind only the porous PLGA scaffold (**Fig. 10.4**). The shapes of the scaffolds made by this technique can be easily changed by using molds of different geometries. The porosity and pore size can also be independently controlled by changing the amount and size of the porogen particles, respectively. It is possible to add other solid materials (e.g., ceramic particles) in the scaffold to improve its mechanical characteristics. Also, there is a possibility to incorporate bioactive agents into such scaffolds. A wide range of substances can be used as scaffold materials, porogens, and molds. It all depends on the purpose of the scaffold to be constructed and on technological preferences. The main disadvantage of this approach is the fact that, in the case of some materials, high melting temperatures during the processing can degrade and inactivate some of the components of the scaffold [4].

Electrospinning uses solutions of polymers, which are ejected across a high voltage potential through a fine orifice, creating very fine fibers. After the evaporation of the solvent, the scaffold made of such fibers has a highly porous structure (**Fig. 10.5**).

Phase separation   In the phase separation approach, upon temperature change initially homogeneous solution becomes thermodynamically unstable and turns into a multiphase system consisting of polymer-rich and polymer-poor clusters. The polymer-rich

Fig. 10.4 Schematics of the melt-molding technique

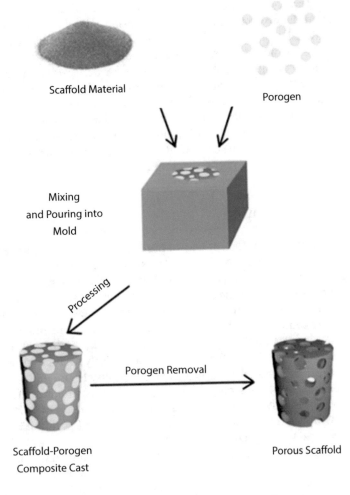

Scaffold Material

Porogen

Mixing and Pouring into Mold

Processing

Porogen Removal

Scaffold-Porogen Composite Cast

Porous Scaffold

Fig. 10.5 Schematics of electrospinning

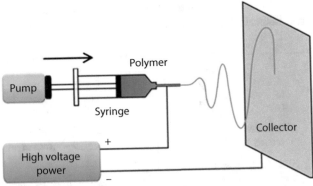

Polymer

Pump

Syringe

Collector

+

High voltage power

−

phase is then turned into a scaffold, while the polymer-poor phase is removed, producing pores. In the case *of freeze-drying*, the polymers/ceramics are emulsified in a solvent, which is followed by pouring this substance into a mold. After this, the solvent is removed to create a porous scaffold. The approach relies on the principle of sublimation.

Textile-based scaffolding uses knitting, weaving, and braiding of polymers to create a defined scaffold structure by using microfibers made from different suitable materials, for example, hyaluronan. Different chemical modifications of initial material allow adjusting water solubility of such fibers from seconds to several months. Fabrics of different shapes, structures, or dimensions can be produced by different textile techniques. Hyaluronan microfibers, for example, are resorbable, implantable, and sterilizable. The size of monofilaments can range from 10 to 200 micron. Multiple types of monofibers can be blended in one yarn from which different patterns of knitted fabric can be then produced. A number of companies, an example being CONTIPRO, can be contacted to purchase or custom-create textile-based materials for tissue engineering applications.

10.4 Bioprinting

Another highly efficient way to produce the desired 3D shape is called additive manufacturing or 3D bioprinting. It includes many approaches, but, in all cases, the general principle is laying down the materials in a layer-by-layer manner by a programmable printing device. 3D printing can create both acellular and cell-seeded scaffolds. In traditional 3D printers, the stage or the extruder moves along linear x-y-z coordinates. In the case of additive lathe 3D printing, bioink is deposited on a rotating mandrel instead of a stage, so the system of coordinates is different (◘ Fig. 10.6). The lathe method is particularly useful to print hollow vessels.

3D printing advantages include the ability to use multiple materials, create complex 3D structures, co-print multiple cell types, and incorporate growth factors and other bioactive molecules in a spatially defined manner. In principle, the 3D printing method is more controllable than other types of scaffolding, and so it can be better tailored to create complex functional tissues or even organs. It is important to realize that today 3D bioprinting is still in its infancy, and numerous difficulties are yet to be overcome in order to print tissues or organs in 3D.

The most popular types of bioprinting include extrusion-based and inkjet-based methods. In *inkjet-based* methods, biomaterials are placed as droplets in a layer-by-layer fashion. These droplets can be formed either by piezoelectric or thermal actua-

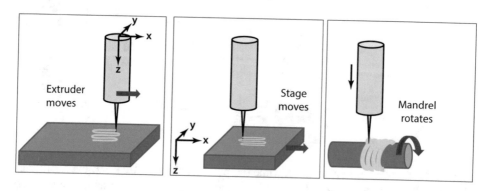

◘ **Fig. 10.6**   Main ways to move 3D printed structure relative to the extruder position

tion. Such printers have high accuracy and high resolution. In the case of *extrusion-based* methods, biomaterials are extruded from the printhead by a pressure gradient, which can be mechanical or pneumatic. This approach is more suitable for cells and bioactive agent incorporation because it does not involve any temperature changes that could harm biological entities.

Depositing material in 3D space represents only the first step in tissue formation. Then cells have to actually connect to each other and start forming the tissue. To help achieve that, current bioprinting protocols rely on biomimicry, autonomous self-assembly, and use of mini-tissue building blocks (also called micromasonry). In *biomimicry* approach, the cells and other components are arranged in a way to be able to mimic a tissue. *Autonomous self-assembly* methods make use of stem cells and cellular embryonic components so that they could organize into tissues themselves. *Mini-tissue fabrication* approach makes the smallest units (structural or functional) of the tissue and then assembles them. These three approaches can be combined in order to create complex tissues or organs.

As noted above, one of the most common 3D bioprinting protocols involves extrusion-based printers that use cell-laden bioinks or bioinks on top of which cells are seeded after printing. The main components of these bioinks have to fulfill three criteria: support cell adhesion, prompt cell-to-cell contact, and cell migration. Successful bioprinting also requires certain geometric accuracy of printed constructs as well as minimal effects of the printing process on cell viability. Specifically, when it comes to cell-laden bioinks, thorough optimization of multiple printing parameters is needed in order to reduce shear stress on cells while producing high-precision 3D constructs. These parameters include the viscosity of bioink material, the gauge of the needle, extruder temperature, and extrusion pressure. For photocurable bioinks, the duration of UV exposure also has to be considered.

Each type of bioink has its advantages and shortcomings. Let's, for example, consider gelatin methacrylate (GelMA) with the notion that many of the considerations below can be applicable to other bioink types. GelMA's main component is gelatin, which enhances tissue regeneration and printability, while the photopolymerizable methacrylamide group allows matrices to be covalently cross-linked by UV light after printing. GelMA is considered mechanically stable and systemically tunable based on its degree of methacrylation. Furthermore, the elastic modulus of GelMA can be adjusted by altering its concentration and printing parameters. In addition, GelMA encapsulates cells, providing a layer of protection during the 3D printing process and associated shear-induced stress. Yet GelMA has a number of shortcomings. Being made from gelatin, GelMA effectively absorbs most of the dyes, including Trypan blue, primary or secondary antibodies. Therefore, immunostaining protocols that work well for monolayer cultures are not effective in staining cells within GelMA constructs even when one increases antibody concentration or incubation time. Another shortcoming is an adverse effect of non-polymerized GelMA+UV initiator mixture on cell viability that occurs when cells are kept within an extrusion syringe for an extended time prior to crosslinking. The latter leads to poor quality of constructs, made toward the end of the printing process. Thus, it is recommended to use the smallest volume possible when printing and make a fresh mixture of cells and GelMA after a minimal number of prints. Lastly, a general assumption that by lowering GelMA concentration cell spreading can be improved doesn't necessarily apply to all cell types. It also has been shown that different cells have different survival rates after 3D printing using GelMA and other bioinks.

Different methods can be used to document the viability of the cells within 3D printed constructs, each having their own advantages and shortcomings. To better understand the effects of 3D printing on cell viability, the release of cytosolic lactate dehydrogenase (LDH assay), fluorescent dye-based LIVE/DEAD assay, and bioluminescence imaging (BLI) can be used. Bioluminescence provides a convenient way to monitor cell growth and proliferation longitudinally and enables to examine the long-term effect of printing steps and not just acute damage to cell membranes [5].

To avoid shear stress associated with bioprinting of cell-containing bioinks, a new 3D bioprinting approach was recently developed (◨ Fig. 10.7, *left panel*). It is called *in-flight droplet mixing* and involves combining droplets of the two gel components in the air. One stream has, for example, fibrinogen with cells and the other thrombin. Because both are in liquid form, minimal stress forces have to be applied to a cell-containing extruder. The droplets become gel particles, which are then deposited using a 3D platform into desired shapes. Another combination can be alginate and calcium solution. This approach also allows printing using higher cell concentrations, which are not attainable in more commonly employed extrusion-based 3D bioprinters.

10.5 Use of Support Bath

Cell-laden bioinks cannot be too stiff without inhibiting cell growth and spreading. This lack of stiffness imposes major limitations on the height, complexity, and stability of printed constructs due to the effects of gravity. Therefore, most reported 3D constructs are only a few millimeters thick and consist of only limited layers of cells. One way to prevent the collapse of macroscopic 3D constructs made from soft biomaterials is to provide a support bath that can hold them in place (◨ Fig. 10.7, *middle and right panels*). Several materials have been shown to be effective as a support bath including hyaluronic acid, carbopol, gelatin-based, and alginate-based slurries. An additional desirable property of support bath material is not to melt in a cell culture incubator. This is because many bioinks that include naturally occurring ECM proteins such as collagen tend to form a more physiologically suitable matrix by slow self-polymerization rather than rapid crosslinking. Therefore, one needs to

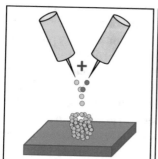

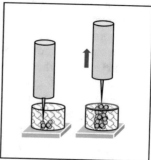

In-flight droplet mixing   Submersion-based   Slurry supported

◨ **Fig. 10.7** Cartoons illustrating principles behind different 3D bioprinting approaches

be able to feed the cells within a 3D printed construct without removing the support bath material while culturing them for extended periods of time. Agarose slurry is one such material, enabling affordable, cell-culture-friendly, and thermally stable support for 3D printed hydrogels [6].

10.6 3D Bioprinters

An increasing number of companies are entering 3D printing market, including BIOX (Cellink), 3DDiscovery EVOLUTION (RegenHU, Switzerland), INVIVO (Rokit Healthcare, South Korea), Allevi (formerly BioBots, USA), Bio V1 (Regemat 3D, Spain), BioScaffolder (GeSIM, Germany), 3D-Bioplotter (EnvisionTEC, Germany), and many others. These newer devices are significantly more advanced compared to the first version of bioprinters. First, there are many more materials that can be used by these machines, including natural and synthetic hydrogels (with or without cells), thermoplastics, decellularized ECM bioinks, conductive inks, or ceramics. Secondly, there is an increase in printer modularity achieved by combining several printing approaches in one device. This way the same machine can use different heads to perform ink-jet, pneumatic, piston-driven, or thermoprinting approaches. In addition, some of these printers also incorporate in their design electrospinning abilities. Another new development is combining microfluidics with 3D printing. This enables on-demand change in bioink formulation *while* printing. Lastly, newer bioprinters strive to become smaller so printing can be done inside a standard biosafety cabinet to preserve sample sterility throughout the printing process and beyond. Achieving such militarization requires the integration of printing parts with the cooling unit, compressor, UV curing, temperature control, and processing module into one device.

Session I

Demonstration

The instructor uses different materials including alginate, gelatin, paraffin, and agarose that can form molds and cast different 3D shapes using biomaterials. The differences between the physical properties of various scaffold materials and shapes that can or cannot be removed without breaking the mold are also discussed.

Homework

Teams are tasked with searching the literature on how to make 3D models of their organ of choice using biocompatible materials.

Session II

Team Exercises

Each team creates a desired 3D shape using biocompatible scaffold material such as alginate or gelatin and matching mold material. Matching pairs of mold and cast material can be either chosen by the students or randomly distributed to the teams.

Homework

Teams prepare PowerPoint slides presentation summarizing the casting process.

ℹ Sample Protocol

Melt molding technique. This protocol is intended to introduce the melt-molding technique to students. The use of alginate as a mold and gelatin as a scaffold material is described. However, it is also possible to do vice versa—to use alginate as a scaffold material and gelatin as a mold. A wide range of materials (such as paraffin, agarose, etc.) can be used as alternative options.

1. Use modeling clay to make a model of desired geometric shape.
2. Choose a dish for molding. Prepare and weigh the appropriate quantity of alginate powder depending on the size of the model and the dish. Pour the powder into the dish.
3. Add distilled water into the dish, at a ratio of 2:1 (water to alginate, v/w).
4. Mix about half a minute. The color of the alginate (Alginate A2010, Alginmax, Major Dental) will turn into pinkish-purple and then return to its original color, which will indicate the initiation of the solidification process.
5. When alginate starts to solidify, immerse the clay model into it (make sure that model shape allows removing it without damaging the mold after the latter solidifies).
6. Carefully remove the clay model without damaging the alginate. After this step, the alginate mold will be ready.
7. Take a heat-resistant glass container. Add gelatin and distilled water in order to make a 25% solution of gelatin.
8. Put the container on an electric stove, and, by constantly stirring, wait until gelatin completely dissolves in water. Do not let the water boil.
9. After the complete dissolution of gelatin, pour the gelatin solution into the mold.
10. Put the dish with the mold and the cast into the freezer and wait until gelatin solidifies completely.
11. After the complete solidification of gelatin, take the dish out of the freezer and carefully, without damaging, detach the gelatin cast object from softer alginate mold.

Take-Home Message/Lessons Learned

After reading this chapter and performing the requested assignments and exercises, students should:

- Gain basic knowledge of different ways how 3D shapes can be made from scaffold material
- Learn about different bioinks used for 3D printing of tissues using cell-containing formulations
- Become aware of adverse effects of 3D printing on cell viability and ways to control it
- Be able to create simple molds and custom casts using available biologically compatible materials

Self-Check Questions

Q.10.1. During the 3D printing process, cells are least likely to be damaged by
   A. UV crosslinking
   B. Shear stress
   C. Lack of oxygen and nutrients
   D. Low temperature

Q.10.2. During 3D printing using pneumatic extrusion, the diameter of extruded lines can be affected by all, EXCEPT
   A. The viscosity of the bioink
   B. The velocity of the needle tip movement
   C. Type of the cells within the bioink
   D. Size of the needle

Q.10.3. The cast of what shape cannot be removed without breaking agarose mold?
   A. Sphere
   B. Cube
   C. Cone
   D. Cylinder

Q.10.4. All of these factors below have been used to help solidify bioink material, EXCEPT
   A. The concentration of calcium ions
   B. Temperature
   C. Duration of UV exposure
   D. Concentration of magnesium ions

Q.10.5. Select the mismatched pair.
   A. Alginate—calcium ions
   B. Gelatin—temperature
   C. Collagen—pH
   D. GelMA—osmolarity

References and Further Reading

1. A. Salerno, G. Cesarelli, P. Pedram, P.A. Netti, Modular strategies to build cell-free and cell-laden scaffolds towards bioengineered tissues and organs. J. Clin. Med. **8**(11), 1816 (2019)
2. P.F. Egan, Integrated design approaches for 3D printed tissue scaffolds: Review and outlook. Materials (Basel) **12** (2019). https://doi.org/10.3390/ma12152355
3. U. Jammalamadaka, K. Tappa, Recent advances in biomaterials for 3D printing and tissue engineering. J. Funct. Biomater. **9**(1), 22 (2018)
4. R.M. Allaf, Melt-molding technologies for 3D scaffold engineering. Funct. 3D Tissue Eng. Scaffolds, 75–100 (2018)
5. P. Koti, N. Muselimyan, E. Mirdamadi, H. Asfour, N.A. Sarvazyan, Use of GelMA for 3D printing of cardiac myocytes and fibroblasts. J. 3D Print. Med (2019). https://doi.org/10.2217/3dp-2018–0017
6. E. Mirdamadi, N. Muselimyan, P. Koti, H. Asfour, N.A. Sarvazyan, Agarose slurry as a support medium for bioprinting and culturing free-standing cell-laden hydrogel constructs. 3D Print. Addit. Manuf. **6**(3), 158–164 (2019)

Bioreactors

Arman Simonyan and Narine Sarvazyan

Contents

© Springer Nature Switzerland AG 2020
N. Sarvazyan (ed.), *Tissue Engineering*, Learning Materials in Biosciences,
https://doi.org/10.1007/978-3-030-39698-5_11

What You will Learn in This Chapter and Associated Exercises
Students will be given basic information about different types of bioreactors that allow
to culture 3D tissue-engineered constructs and the ways to monitor cells within them.
During the practical session, students will be putting together a simple bioreactor of
their choice.

11.1 Basic Types of Bioreactors

A bioreactor is a manufactured or engineered device that supports a biologically
active environment. In the context of tissue engineering, bioreactors are used to support cell growth in the decellularized tissue, artificially created recellularized scaffold,
or any other type of biocompatible matrix. Several operational conditions within the
reactor can be modified and controlled. They include pH, temperature, oxygen tension, and the rate of media perfusion as well as the ability to apply external stimuli
such as mechanical forces or electrical stimulation. The key principles to be employed
in any bioreactor are as follows:
- Simplicity and quickness of assembly
- Efficiency of tissue formation in a short span of time
- Sterility
- Non-toxicity of materials of assembly
- Ease of sterilization for repeated use

The specific design of bioreactors varies depending on the targeted tissue. For engineered cardiac tissue, for example, a bioreactor can have electrodes in order to pace
the tissue. For bone or cartilage, it has to provide a specified degree of compression,
preferably in a time-cyclic fashion. Below we briefly discuss several types of bioreactors based on a comprehensive review by Martin et al. [1].

I. *Spinner flask bioreactor.* A spinner flask bioreactor is the most basic and simple
    form of bioreactor used in tissue engineering. It assumes the fixed position of
    the scaffold in the flask, while the magnetic stirrer constantly mixes the media
    around (◻ Fig. 11.1). This enables the constant flow of nutrients and oxygen
    through the scaffold. Typically, spinner flasks are around 120 mL in volume
    and run at 50–80 rpm, and 50% of the medium used in them is changed every
    2–3 days. These types of bioreactors are usually used for the recellularization
    of tissues with a high value of surface-to-volume ratio, that is, for relatively
    small or "flat" scaffolds. For example, spinner-flask bioreactors have been suc-
    cessfully used in articular cartilage production in vitro.

II. *Rotating wall bioreactor.* Rotating wall bioreactor consists of inner and outer
    cylinders whose walls are rotating at a constant speed (◻ Fig. 11.2). It is
    commonly used when there is a need to reduce shear stress since cells here
    grow in a microgravity environment. The scaffold in this bioreactor is not
    fixed; it moves freely in the media. Media can be exchanged by stopping the
    rotation temporarily or by adding a fluid using a pump, whereby media are
    constantly pumped through the vessel. Gas exchange occurs through a gas
    exchange membrane, and the bioreactor is typically rotated at speeds of
    15–30 rpm [2].

Bioreactors

■ Fig. 11.1 Spinner flask bioreactor

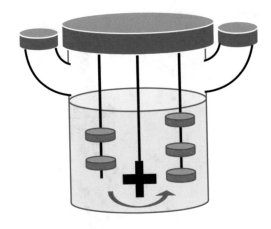

■ Fig. 11.2 Rotating wall bioreactor

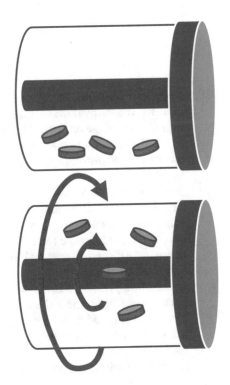

Fig. 11.3 Perfusion-based bioreactor

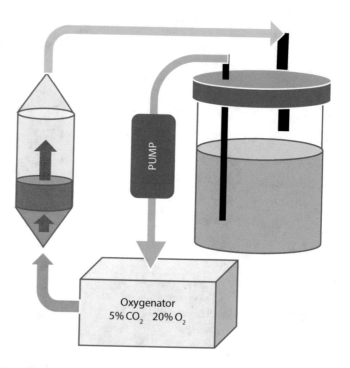

III. *Compression bioreactor.* As the name implies, compression bioreactors are used for the cultivation of tissues that need pressure in order to develop, for example, cartilage or bone. The compression bioreactor consists of a chamber where the scaffold is placed, a motor, a system providing linear motion, and one or several pistons applying static or dynamic compression loads upon the tissue [3].

IV. *Strain bioreactor.* Strain bioreactors are very similar to the compression systems in their structure and mechanism. However, instead of flat plates as in a compression bioreactor, a way of clamping the scaffold into the device is needed so that a tensile force can be applied [4]. These systems have been used to engineer diverse tissues including tendon, bone, cartilage, ligament, cardiac, and vascular tissues.

V. *Hydrostatic pressure bioreactors.* Hydrostatic pressure system normally consists of a chamber where the scaffold is placed, a pressure-applying means (e.g., a piston controlled by an actuator), a water-filled pressure chamber controlled using a variable backpressure valve, and an actuator. The pressure is created via an impermeable membrane to assure sterility. HP bioreactors are used, for example, to mimic the conditions of intervertebral discs [5].

VI. *Flow perfusion bioreactor.* Perfusion systems typically consist of a chamber where the scaffold is placed, a liquid reservoir chamber, a pump, and connecting tubes (**Fig. 11.3**). The cell-containing suspension is continuously pumped through the scaffold providing optimal cell attachment and growth. Flow perfusion bioreactors, in contrast to the spinner flask or rotating wall bioreactors, solve the problems caused by the static composition of media.

The constant flow in perfusion systems not only combats the problem of the cell concentration on the exterior of a scaffold caused by static media but also provides a constant waste elimination and satisfies nutrient inflow necessity. Furthermore, the media can easily be replaced in the media reservoir. The perfusion systems have been shown to result in more homogeneous cell distribution compared to the static media systems [6]. One of the disadvantages of flow perfusion bioreactor is the alignment of the cells parallel to the scaffold lining, while some systems (e.g., articular cartilage) would require perpendicular alignment.

VII. *Microfluidics-based bioreactors.* To monitor and record fluorescent signals from live cells within tissue constructs new types of bioreactors are being developed. They enable perfusion of small pieces of tissue placed inside chambers with #1 coverslip glass bottoms. One such example is a miniaturized optically accessible bioreactor (MOAB), which has several optically clear chambers with magnetically attachable lids connected to a microfluidic device that controls the flow. The device allows keeping cell constructs perfused with cell culture media while doing real-time microscopy. The chambers can also be connected to each other to mimic paracrine signaling (paracrine signaling involves substances released from neighboring cells or tissues).

VIII. *In vivo bioreactors.* One of the biggest challenges of bioreactors is the long-term maintenance of cells within scaffold, in terms of nutrition and oxygen delivery. Within the body, most cells are found no more than 100–200 µm from the nearest capillary, which provides them with necessary substances and eliminates the waste. In vitro, the effective delivery mechanism remains a limiting factor. Therefore, in most cases of tissue-engineered organs, oxygen delivery relies merely on surface diffusion, resulting in viable tissues only a few hundred microns in thickness. Several ways to address this problem are being developed. They are shown in a cartoon form in ◘ Fig. 11.4. based on the article by Lovett et al. [7]. Current approaches include incorporation of angiogenic factors into scaffold material, addition of endothelial cells to generate capillary-like sprouts that can then connect to the vessels of the host, use of perfusion bioreactors that increase passive flow through the tissue, integration of microfluidic channels made from synthetic or natural polymers, insertion of pre-formed channels from endothelial cells into hydrogel scaffolds, in vivo vascularization of multilayered cell sheets followed by addition of more layers, and finally printing of vessels within the rest of 3D printed tissue. By combining these approaches, recently it became possible to create dense tissues up to 1 cm thick [8]. Newer materials and approaches to aid in vascularization of engineered tissues are continuously being developed, and students are encouraged to read recent articles and reviews on this subject [9–11].

As for now at least, the most efficient way to solve the vascularization problem is to implant tissue constructs into a living organism so it can be fed by the host blood vessels. In time, smaller vessels sprout into the construct forming a new vasculature. This approach essentially uses "in vivo bioreactor" with the host body taking over the vascularization process. When vascularized constructs reach maturity, they can be re-implanted to their final implantation site or even a new host.

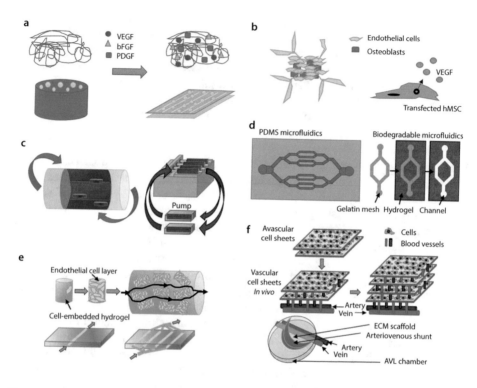

Fig. 11.4 Schematic diagrams of different scaffold vascularization approaches. **a** Incorporation of angiogenic factors, **b** addition of endothelial cells to generate capillary-like sprouts, **c** use of perfusion bioreactors, **d** integration of microfluidic channels made from synthetic or natural polymers, **e** insertion of preformed channels from endothelial cells into hydrogel scaffolds, **f** in vivo vascularization of multi-layered cell sheets followed by addition of more layers

11.2 How to Monitor Cells Within a Scaffold

Tissue assays described in ▶ Chap. 5 can also be used to evaluate either viability or specific activity of the cells within scaffold material. In addition, as briefly described in ▶ Chap. 7, confocal microscopy enables visualization of the cells labeled with fluorescent markers through the thickness of the sample (to about 100–200 micron depth). Examining different layers of the tissue allows for the confirmation of the presence of viable cells within the whole volume of the recellularized tissue or scaffold and provides 3D information for data representation.

Another way to image live constructs without fixing them is to use bioluminescence. Bioluminescence detects photons emitted by an enzymatic reaction in which a luciferase oxidizes a substrate. Bioluminescence imaging has been used extensively for tracking live cells within a living organism, giving real-time information on their survival, growth, and proliferation. It requires additional steps such as the introduction of the luciferase gene into the target cell genome and injection of luciferin. Depending on the intensity of the luminescence signal obtained, the time needed to acquire an image may be as little as 1 s or as long as 10–20 min.

Session I

Demonstration

The instructor builds a simple bioreactor using available tools. It is important to demonstrate the simplicity and minimalism in construction. Yet, being simply built, the bioreactor still needs to support all the features listed in the section above: non-toxicity of material, ease of sterilization, capacity to exchange media for nutrient exchange as well as being compatible with cell incubator conditions including high humidity and 37 °C (for example one needs to consider that metal parts can rust).

Homework

Teams are tasked with searching the literature to find design of a feasible bioreactor suitable for long-term culturing of their organ of choice.

Session II

Team Exercises

Each team builds a different type of bioreactor using available tools. The type of bioreactor can match the tissue that the team is planning to cultivate. The practice will help students to build a more elaborated device for the actual recellularization.

Homework

Team members meet to decide how they will proceed with engineering the organ of their choice within the next 2–3 weeks. Methods, hypotheses to be tested, animal species from which to obtain organs and cells, required reagents, and assignment of individual tasks—all these items must be debated and presented to the instructor for further discussion. Each team then meets with the instructor individually to discuss the design of their planned experiments.

11.3 Histology of 3D Engineered Tissue

Histological techniques are the traditional methods employed for the assessment of tissue structure. They can be used to confirm the presence of cells in 3D scaffolds by analysis of two-dimensional slices prepared from tissue samples. Using the proper staining, it's possible to identify the specific types of cells in the sample. By analyzing multiple histological sections from the same sample, the results are often extrapolated to the three-dimensional structure. Measuring histomorphometric parameters such as percent of the viable cells yields quantitative results. Histology, in contrast to the confocal imaging or other types of live imaging, is a destructive technique that does not allow further analysis of the samples as cells are fixed.

ⓘ Sample Protocols

Static Media Bioreactor 1

1. Pick a suitable flask/bottle with a cap to keep the scaffold in.
2. Place the scaffold in.
3. Fill the flask with media.
4. Place the flask into the thermoshaker and adjust the temperature and shaking motion.
5. Stop the shaker and change the media when required.

Static Media Bioreactor 2

1. Pick a suitable flask/bottle with a cap to keep the scaffold in.
2. Put in a magnet.
3. Affix a porous membrane (any round-shaped material stiff enough to hold a scaffold and permeable for media) inside the flask (4–5 cm above the floor) so that the flask is divided into lower and upper parts.
4. Place the scaffold on the membrane.
5. Fill the flask with media.
6. Place the flask on the magnetic stirrer and turn it on (adjust the motion).
7. Put the whole construction in the cell incubator with adjusted temperature.
8. Stop the device, take it out, and change the media when required.

Flow Perfusion Bioreactor

1. Pick a suitable flask/plate/bottle with a cap and place the scaffold in side.
2. Pick a suitable container and fill it with media.
3. Fix one lab hose going from the media container into the scaffold flask and another one from the flask to the container.
4. If the scaffold has vasculature, adjust the corresponding edges of hoses in a way that they carry media into the incoming vessel and take the media out from the outgoing vessel (e.g., fix the pipette tips on the hoses' edges and insert them into the corresponding vessel entrances).
5. Connect pumps to the hoses so that they push the liquid in corresponding directions.
6. Put the whole construction into the cell incubator with adjusted temperature.
7. Change the media regularly.

Take-Home Message/Lessons Learned

After reading this chapter and performing the requested assignments and exercises, students should:

- Understand the basic principles that any type of bioreactor must comply with
- Be familiar with the main categories of bioreactors and types of engineered tissue they are designed for
- Be aware of different ways to vascularize engineered tissue
- Be capable of building a simple bioreactor using commonly available plasticware and tubing.

Self-Check Questions

Q.11.1. Select the best type of reactor for a tissue-engineered vessel.
A. Rotating wall bioreactor
B. Flow perfusion bioreactor
C. Compression bioreactor
D. Spinner flask

Q.11.2. To recreate in vivo environment while growing engineered heart constructs, a bioreactor's design should include
A. Electrical stimulation and cyclical stretch
B. Daily media change
C. Highly periodic changes in media calcium concentrations
D. Mechanical tissue compression

Q.11.3. To recreate in vivo environment while growing engineered bone constructs, a bioreactor's design should include
A. Electrical stimulation and cyclical stretch
B. Daily media change
C. Highly periodic changes in media calcium concentrations
D. Mechanical tissue compression

Q.11.4. Tissue vascularization strategies include
A. In vivo vascularization of sequential layers of engineered tissue
B. Creation of small artificial channels using microfluidics
C. Incorporation of angiogenic factors stimulating endothelial cell proliferation
D. All of the above

Q.11.5. General bioreactor design must enable a user to do all of the following, EXCEPT the ability to _____ which is optional.
A. Regulate temperature and oxygen delivery
B. Control rate of media perfusion
C. Observe cells while they are forming tissue
D. Sterilize bioreactor for repeated use

References and Further Reading

1. I. Martin, D. Wendt, M. Heberer, R. Langer, E. Al, The role of bioreactors in tissue engineering. Trends Biotechnol. **22**(2), 80–86 (2004)
2. L.E. Freed, G. Vunjak-Novakovic, Microgravity tissue engineering. Vitr. Cell Dev. Biol. Anim. 33(5), 381–385 (1997)
3. M. Sladkova, G. de Peppo, Bioreactor systems for human bone tissue engineering. Processes **2**(2), 494–525 (2014)
4. N. Plunkett, F.J. O'Brien, IV.3. Bioreactors in tissue engineering. Stud. Health Technol. Inform. **152**, 214–230 (2010)

5. J. Zvicer, B. Obradovic, Bioreactors with hydrostatic pressures imitating physiological environments in intervertebral discs. J. Tissue Eng. Regen. Med. **12**(2), 529–545 (2018)
6. J. Glowacki, S. Mizuno, J.S. Greenberger, Perfusion enhances functions of bone marrow stromal cells in three-dimensional culture. Cell Transplant. **7**(3), 319–326 (1998)
7. M. Lovett, K. Lee, A. Edwards, D.L. Kaplan, Vascularization strategies for tissue engineering. Tissue Eng. Part B Rev. **15**(3), 353–370 (2009)
8. D.B. Kolesky, K.A. Homan, M.A. Skylar-Scott, J.A. Lewis, Three-dimensional bioprinting of thick vascularized tissues. Proc. Natl. Acad. Sci. **113**(12), 3179–3184 (2016)
9. M.D. Sarker, S. Naghieh, N.K. Sharma, X. Chen, 3D biofabrication of vascular networks for tissue regeneration: A report on recent advances. J. Pharm. Anal. **8**(5), 277–296 (2018)
10. D. Richards, J. Jia, M. Yost, R. Markwald, Y. Mei, 3D bioprinting for vascularized tissue fabrication. Ann. Biomed. Eng. **45**, 132–147 (2017)
11. L.A. Herron, C.S. Hansen, H.E. Abaci, Engineering tissue-specific blood vessels. Bioeng. Transl. Med. **4**(3) (2019)

1

Reporting Results

Narine Sarvazyan

Contents

© Springer Nature Switzerland AG 2020
N. Sarvazyan (ed.), *Tissue Engineering*, Learning Materials in Biosciences,
https://doi.org/10.1007/978-3-030-39698-5_12

What You Will Learn in This Chapter and Associated Exercises
Students will learn how to make effective poster and oral presentations using data generated during previous weeks.

12.1 Final Student Projects

At the end of the 3-month period, students are given 3 weeks to conduct their own experiments to engineer their tissue of choice. Students are first asked to search the literature and present the most suitable published protocols. This is followed by an in-class discussion to determine how feasible it is to perform these experiments both time and reagent wise. The availability of live animals or freshly excised tissue has to be established ahead of time. It is also important to discuss what exactly will be measured or imaged to document the creation of teams' tissue of choice. Students are free to offer up to ten options of what they want to measure. After the discussion, two to three of the most suitable options can be finally chosen. Students are encouraged to test at least one specific hypothesis within the framework of their experiments. This can be the effect of animal age, type of species, bioreactor type, etc. Team members then work together using their own schedule to create and examine their tissue-engineered constructs. It is important to understand that the gained data will be later used in the team's poster/oral presentations; so pictures should be captured under the same conditions to show the representable results.

12.2 Poster Preparation

At the end of the 3-week period, teams are asked to prepare posters that summarize the results of their efforts. Due to limited course time, participants are advised to begin poster preparation even before they collect all the data, by starting to formulate their introduction, methods, and hypothesis sections. Each poster must be formatted according to the standard guidelines and include Abstract, Introduction, Methods, and Results sections. The latter usually contains four to eight figures with figure legends followed by the Conclusion section. Acknowledgments and References sections are included at the very end. The title should be concise, not very specific, and no longer than two lines. The authors, their affiliation, and the place where experiments were performed have to be mentioned below the title.

Abstract section   An abstract is a condensed description of the performed work so viewers can quickly catch the general idea of the study. To formulate an abstract, students are advised to compose one to three sentences each describing study rationale, methods, results, and conclusion. These sentences can be then put together into a coherent text.

Introduction section   Through the Introduction section, the key points leading to the study should be delivered—why this particular tissue was chosen, what was done by others in this field, why the selected features were measured, and why the work was important clinically. Be attentive not to repeat the abstract content in the introduction. The use of flow charts and any other visual effects in this section is encouraged.

Methods and results sections   Methods can be described only briefly, particularly if there is a lack of space. Most space should be given to the results. The charts, graphics, and images must be clear, well-labeled, and easily understood. Each documented step in the experimental protocol or measured result can be presented as a separate figure. It is advised to make line art for the graphs in dark colors while keeping the background white or light color. This will make it easier to use the created graphs for other types of publications. Each image needs to have a scale bar. Each numerical value has to be accompanied by statistical information.

Conclusion section   This section should summarize results and whether collected data supported or disproved the authors' initial hypothesis. Future directions and ideas can also be mentioned in the conclusion. The Reference section is optional for posters. However, it is advised to include the most critical papers on which the study was based. Giving the acknowledgments to people, who supported and helped during the whole project, is also important. It is best to mention the type of specific assistance and not simply list the names.

Overall design   Students are advised to use the maximal contrast between the font color and the poster background, to minimize unnecessary details, and to use cartoons to better portray experimental procedures. It is important that the poster content is visible from a distance of approximately 2 m. The quality of the images must be high. The font of the text should be large enough and appropriate to the poster size. Two to three colors can be used to make fonts or frames of the poster more eye-catching. Using more colors can actually distract from the poster's main content. Any additional artistic effects (i.e., 3D graphs, shadows, etc.) should be carefully considered for them not to divert the viewer from the main message. A scientific poster should visually attractive, but its main point is to present the data to the viewer. PowerPoint, Keynote or Adobe Photoshop programs can be used as tools to create the poster. It is advised to convert the file into PDF format to see how the final version of the poster looks like when printed.

12.3 Oral Presentations

The last part of the course is the teams' oral presentations. This can be organized as an event open to the interested public. After the course director introduces the main subject of the course, teams take turns presenting their projects. Students are encouraged to be creative about the exact format of their oral presentation while touching on the main points These points include why and where their tissue of choice can be clinically useful, what has been accomplished by others to build it, what were the teams' methods, what worked or did not work, and what was the most exciting part of the experiments conducted. Depending on the audience, the presentation can be more public-oriented and less scientifically specific. A slide-based presentation is probably the most practical option. The core of a good presentation is its visual representativeness. Only high-quality pictures should be included. The number of slides must be limited to one slide per minute. The mean duration of such type of presentation can range from 15 to 20 min. The structure of the presentation is similar to the poster structure but with fewer details. The overall appearance of slides is very important as it helps to keep the audience's attention. Each slide should not contain too much text. As much as possible, the text should be replaced with cartoons, pictures, charts, graphs, and tables. Students can use the background material they prepared for earlier sessions of the course. Sharing a few funny moments the team experienced during the course allows connecting better with the audience. The talk of each student should be well prepared and coordinated with other teammates. Students should be ready to answer questions from the audience following their presentation.

12.4 Peer-to-Peer Ratings and Student Feedback

Practicing oral and poster presentations a day or two before the public presentation is highly recommended. Such a preview session gives the course director an excellent opportunity to conduct peer-to-peer review and rating of the team's presentations. The latter process has multiple benefits. First is immediate feedback that can improve the quality of a team's oral presentation. Secondly, it provides the instructor with the opportunity to discuss mistakes made by individual teams and turn them into learning opportunities for the entire class. Thirdly, students learn how to provide constructive criticism to fellow students. Finally, collected peer-based ratings can be used to issue awards at the completion of the course. Awards for the best poster, the best oral presentation, the most helpful member of the team, the least contaminated cultures, or any other additional/humorous achievements can be given to the teams as well as individual participants. Students are also asked to complete course evaluation forms.

Take-Home Message/Lessons Learned

After reading this chapter and preparing their own presentations, students should:
- Be able to analyze and present their own data
- Be able to use appropriate fonts and colors to effectively present their data
- Understand the general structure of either poster or oral presentation and contents of its main elements
- Know how to engage in peer-review grading of presentations

Self-Check Questions

Q.12.1. The main goal of the poster presentation is to
A. Present all the data in detailed format
B. Produce stunning visual effects
C. Include all relevant references
D. Outline results and conclusions in a visually clear way

Q.12.2. The recommended number of slides for oral presentation is
A. One slide per minute
B. One slide per 5 minutes
C. Three slides per minute
D. Ten slides per minute

? Q.12.3.   In order to use the same figures for other types of publications, it is recommended to format poster figures using
A. Light colors on a dark background
B. Dark colors on a light background
C. A close match between background color and graphics
D. Red color on yellow background

? Q.12.4.   Choose the correct statement.
A. Humor is not allowed in an oral presentation.
B. Poster content should be visible from approximately 2 m.
C. Oral presentation needs to include all methodological details.
D. Acknowledgment section should only include funding sources and not the names of the people who assisted authors with their studies.

? Q.12.5.   Most space in a typical science poster has to be given to the _____ section.
A. Introduction
B. Methods
C. Results
D. Conclusion

Supplementary Information

Answers to Self-Check Questions – 144

Answers to Self-Check Questions

- Chapter 1 Answers: 1A, 2B, 3D, 4B, 5A
- Chapter 2 Answers: 1D, 2C, 3A, 4E, 5B
- Chapter 3 Answers: 1C, 2B, 3D, 4C, 5B
- Chapter 4 Answers: 1B, 2C, 3B, 4A, 5C
- Chapter 5 Answers: 1C, 2C, 3B, 4B, 5D
- Chapter 6 Answers: 1D, 2B, 3C, 4D, 5C
- Chapter 7 Answers: 1D, 2A, 3D, 4B, 5D
- Chapter 8 Answers: 1A, 2C, 3B, 4A, 5B
- Chapter 9 Answers: 1A, 2D, 3C, 4C, 5B
- Chapter 10 Answers: 1D, 2C, 3A, 4D, 5D
- Chapter 11 Answers: 1B, 2A, 3D, 4D, 5C
- Chapter 12 Answers: 1D, 2A, 3B, 4B, 5C

Chapter 1 Answers

Q.1.1. **The correct answer is A.** *Google Scholar* is a comprehensive free online portal that enables search of scientific literature including links to previous papers that cited a particular article. Other choices are incorrect. *ImageJ* is a software to analyze images; *SciHub* is a pirate website to access research papers; *JoVE* is a video methods journal with restricted access.

Q.1.2. **The correct answer is B.** *PubMed* is a free online search engine encompassing over 30 million citations for biomedical literature from MEDLINE, life sciences journals, and online books. Once article of interest is found, the tab on the right suggests multiple articles on the similar subject. Other choices are incorrect. Although *Google Scholar* also has a tab called "Similar Articles," its algorithm is currently is not performing as well as the one employed by PubMed platform. *ImageJ* is a software to analyze images; it is not a search engine. *SciHub* does not have an option of referring to similar articles.

Q.1.3. **The correct answer is D.** Full content of patents with detailed protocols is searchable using Lens.org portal. It also has an extensive array of visualization tools. Other choices are incorrect. *Google Scholar* and *PubMed* do provide links to the articles, but not all of them have free access; therefore, the Methods section can be unavailable. *ImageJ* is a software to analyze images; it is not a search engine.

Q.1.4. **The correct answer is B.** *ImageJ* is a free online software to process, edit, and analyze images. The rest of the options refer to sites to search scientific literature.

✅ Q.1.5. **The correct answer is A.** Standard error of the mean (SEM) is calculated as the sample standard deviation (SD) divided by a square root of the number of samples. Assuming at least two measurements or samples are being considered, the SEM will be always smaller than SD.

Chapter 2 Answers

✅ Q.2.1. **The correct answer is D.** Access to organ vasculature varies and can include artery (i.e., in case of the kidney) or vein (i.e., in case of the liver). Therefore, the correct answer is "depends on an organ."

✅ Q.2.2. **The correct answer is C.** Capillary diameter varies between 5 and 15 micron—just enough for red blood cells to squeeze through. The diameter of 1 mm will correspond to a small artery or vein. Diameter of 100 micron is an appropriate estimation for an arteriole or a venule. Diameter of 1 micron is incorrect as it will not allow red blood cells to pass through. These cells are about 7 microns wide. They can adjust their shape a bit by folding, but not seven times.

✅ Q.2.3. **The correct answer is A.** Collagenase breaks peptide bonds in collagen, one of the main extracellular matrix proteins holding cells together. Collagenase preparations also have additional proteolytic enzymes aiding in the release of cells from the extracellular matrix. Other choices are incorrect. Heparin is used to prevent blood clotting. Detergents will release cells but will cause irreversible cell damage by dissolving their membranes. Since calcium ions are involved in multiple types of cell-to-cell junctions, low and not high calcium concentration solutions can help to dissociate cells. Ketamine/xylazine is the most common anesthetic combination for rodents and other animals.

✅ Q.2.4. **The correct answer is E.** Ketamine/xylazine is the most common anesthetic combination used to anesthetize for rodents and other animals. Other choices are incorrect. Collagenase is an enzyme that digests collagen. Heparin is a drug against blood clotting. Use of detergents will destroy cell membranes besides dissociating cells from each other. Solutions with low (not high) concentrations of calcium can be useful for separating cells.

✅ Q.2.5. **The correct answer is B.** Heparin is an anticoagulant that prevents the formation of blood clots. Other choices are incorrect. Collagenase dissolves collagen by breaking its peptide bonds. Detergents will irreversibly damage cell membranes. High calcium concentration solution won't help to dissociate cells. Ketamine/xylazine is an anesthetic combination of choice for rodents and other animals.

Chapter 3 Answers

✓ Q.3.1.   **The correct answer is C.** Cadherin family of proteins links cells together. These proteins are found on cell membranes, not in ground substance. Other choices are incorrect since proteoglycans, glycosaminoglycans, and glycoproteins are all ingredients of ground substance, the latter being extracellular matrix without fibrous materials such as collagen and elastin.

✓ Q.3.2.   **The correct answer is B.** Communicating junctions, that is, gap junctions, allow molecules less than ~500 Da to diffuse in and out of the cells. A is incorrect because the main role of *occluding junctions*, also called tight junctions, it to prevent leakage of fluids through paracellular pathways. C is incorrect since *anchoring junctions* are involved in cell-to-cell attachment, recognition, and morphogenesis, and not involved in sealing off the fluid passage. D is incorrect since *selectins* are mostly found in white blood cells (L-selectin), endothelial cells (E-selectin), and platelets (P-selectin).

✓ Q.3.3.   **The correct answer is D.** The role of ECM includes all the functions listed in A–C; therefore, D is correct.

✓ Q.3.4.   **The correct answer is C.** This is because collagen and elastin are large structural fibers present in the extracellular matrix, while immunoglobulins are cell surface and soluble proteins involved in inflammation and immune response. A, B, and D are incorrect since cadherin, selectin, and integrin belong to the same class of *cell adhesion proteins*; fibronectin, laminin, and osteopontin belong to the same class of attachment *glycoproteins*; while dermatan sulfate, heparin, and hyaluronan are members of *glycosaminoglycans* family.

✓ Q.3.5.   **The correct answer is B.** Coating cell plates with albumin *prevents* cell attachment. Other choices are incorrect because collagen, polylysine, and laminin are among the most commonly used molecules to make the surface of glass or plastic plates suitable for cells to attach and grow.

Chapter 4 Answers

✓ Q.4.1.   **The correct answer is B.** Calcium is a key co-factor for multiple cell-to-cell adhesion proteins; therefore, most protocols call for lowering calcium concentration to ease cell dissociation. Other choices are incorrect since although sodium, potassium, and magnesium are essential in cell function and homeostasis, they are not directly involved in cell-to-cell adhesion.

✓ Q.4.2.   **The correct answer is C.** ATPases are family of key enzymes responsible for splitting ATP, the main "energy currency" of living tissues, for ion transport and numerous chemical reactions inside the cell. The addition of ATPase to the solution will not aid the release of the cells from the tissue. Other choices are incorrect. Collagenase and trypsin are the most common enzymes to digest connective tissue holding cells together. DNAse is commonly used to digest DNA molecules released from damaged cells into smaller pieces. Otherwise, long strains of DNA can trap the remaining intact cells interfering with isolation steps.

Q.4.3. **The correct answer is B.** Isopycnic centrifugation enables to separate cells based on their density by spinning cells through a density gradient; therefore, cells accumulate in the layer with density identical to their own. A is incorrect since flow cytometry separates cells based on their shape, size, and mass (in case of unstained cells). C is incorrect as Trypan blue staining allows to distinguish between live and dead cells, not their density. D is incorrect as Coulter counter helps to count large quantities of isolated cells and not separate them based on density.

Q.4.4. **The correct answer is A.** Trypan blue staining allows to distinguish between live and dead cells, and not to separate them based on their size. All the other three methods can be used to estimate particle sizes.

Q.4.5. **The correct answer is C.** It is critical to collect samples at a specific time when performing differential centrifugation. If a sample is centrifuged for too long, all particles will end up being in the pellet. It is less critical to collect samples at a specific time when using either step or isopycnic centrifugation as particles will be accumulated at the boundary between low and high densities regardless of centrifugation time.

Chapter 5 Answers

Q.5.1. **The correct answer is C.** Chondroitin is produced by chondrocytes serving as a functional, tissue-specific assay. Other options are incorrect. LDH release, reduction of resazurin, and permeability to Trypan blue permeability are used to evaluate cell viability regardless of the type of cells or tissue.

Q.5.2. **The correct answer is C.** Bioluminescence is commonly employed for non-invasive monitoring of implanted tissues. Choices A and B are incorrect as tensile and load-bearing assays are usually used for ex vivo examination of bone and cartilage constructs. Choice D is incorrect since tissues have to be excised to be stained with LIVE/DEAD dyes, followed by imaging under a fluorescent microscope. This assay can be also used to evaluate the viability of cultured cells or tissue constructs as it labels cytosol of viable cells green while nuclei of dead cells stain red.

Q.5.3. **The correct answer is B.** The term "neutral control" is not being used. Choice A (Blank) usually stands for a sample without assay reagents; choice C (Positive control) stands for a sample with a known response to the drug or treatment; choice D (Negative control) stands for a sample without any treatment or treated with a compound that has an opposite effect.

✓ Q.5.4.   **The correct answer is B**. LDH is released into the media and needs to be transferred into another vessel or a new well with a defined amount of fluorescent
substrate (i.e., NADH) for the reaction to take place. Other choices are incorrect since the conversion of resazurin, MTT, and ethidium bromide can be
monitored while the sample remains in the same well.

✓ Q.5.5.   **The correct answer is D**. DNA mutation rate is used to evaluate exposure
to radiation, free radicals, or cytotoxic drugs but is not considered to be
a metabolic marker. Choices A–C are incorrect. They all reflect different
aspects of cell metabolism.

Chapter 6 Answers

✓ Q.6.1.   **The correct answer is D**. Cell cultures, at least for now, are not classified as
synthetic or natural. Choices A–C are incorrect. Cell cultures can be either
suspension or adherent, finite or continuous, plant or animal.

✓ Q.6.2.   **The correct answer is B**. Life evolved in sea and sodium ions are the main ions
of the extracellular space. A solution of 0.9% NaCl creates a physiological
osmolarity of about 300 mOsm and serves as the main component of cell culture and washing media. High extracellular sodium is also required for maintaining cell resting potential. Choice A is incorrect. If physiological values of
osmolarity are to be created using potassium salts, it will abolish resting cell
membrane potential. Choices B and D are incorrect. Calcium and magnesium
ions are needed for normal cell function, but only in small amounts. They can
be omitted for short-term storage or handling of cell cultures.

✓ Q.6.3.   **The correct answer is C**. Antibodies are used for immunostaining or immu-
noprecipitation protocols and not during cell culturing. Choices A, B, and
D are incorrect since all media formulations include some type of buffer and
nutrients. Antibiotics are commonly added to most cell culture medium formulations, except specific cases—an example being direct cell differentiation
protocols.

✓ Q.6.4.   **The correct answer is D**. Nitrogen levels are usually not controlled as nitrogen is an inert gas. Other choices are incorrect. Typical cell culture incubators are designed to control near-physiological temperature (35–37 °C) and
carbon dioxide levels (4–7%). When cultures of cells from insects of cold-
blooded animals are used, the temperature can be set at a different value.
Humidity is usually maintained at near 90–95% by placing a tray filled with
water inside the incubator.

✓ Q.6.5.   **The correct answer is C**. Physiological levels of pH are within 7.2–7.4, and,
with rare exceptions, media formulations are designed to have pH in this
range. Choice A is incorrect since levels of carbon dioxide have a direct
impact on media pH. When dissolved, carbon dioxide becomes carbonic
acid, making media more acidic. Choice B is incorrect. Unless humidity is

so low that water starts to evaporate from the cell culture media, the impact of humidity levels on pH is negligible. Choice D is incorrect. Buffers are very important to balance medium pH. Sodium bicarbonate is added to counteract carbonic acid from dissolved carbon dioxide. Phosphate and buffer are also commonly included in most media formulations.

Chapter 7 Answers

✅ **Q.7.1.** **The correct answer is D.** Common binocular microscopes have 10× ocular lens. When using 100× objective, 1–2 micron objects will be seen by a naked eye magnified 1000 times (10 × 100) or about 1–2 mm. Other objective choices are incorrect as mitochondria will be too small to visualize.

✅ **Q.7.2.** **The correct answer is A.** Common binocular microscopes have 10× ocular lens. When using 2× objectives, 2 × 2 mm object will be seen by a naked eye magnified 20 times (2 × 10) or about 4 × 4 cm. Other choices are incorrect as the image of described piece of an engineered tissue will be too large to fit into a single field of view.

✅ **Q.7.3.** **The correct answer is D.** In the case of a single excitation source, the most optimal pair is TRITC and 7-AAD. Both can be excited by a light source ~550 nm while having emission peaks separated by over 50 nm (~580 nm for TRITC and ~650 nm for 7-AAD). Other choices are incorrect. All other pairs have separate excitation peaks. FITC and TO-PRO-3 are excited by ~500 nm and ~650 nm, FITC and 7-AAD by ~500 nm and ~550 nm, and TRITC and TO-PRO-3 by ~550 nm and ~650 nm, respectively.

✅ **Q.7.4.** **The correct answer is B.** Based on 7-AAD spectra, the following set of filters—540–570 nm, 600 nm, 620 nm long-pass for excitation/dichroic/emission filter, respectively—should be most suitable. Choice A is incorrect as dye excitation will be minimal at 480 nm. Choice C is incorrect since the dichroic mirror has to separate excitation and emission ranges. Case D is incorrect since 600 nm short-pass will cut almost all emitted photons.

✅ **Q.7.5.** **The correct answer is D.** Label ∞/0.17 on the external shield means that the surface of a sample has to be <170 microns away from the objective lens. Choice A is incorrect because confocal microscopes use photons, not beams of electrons, to create crisp images of tissue at different depths. Choice B is incorrect because frozen Mowiol aliquots can be stored for extended periods of time. Choice C is incorrect since the use of acetoxymethyl ester AM enables the dye to enter and become trapped inside the cell. Therefore, it is not used for storing stained cells.

Chapter 8 Answers

✅ Q.8.1. **The correct answer is A.** Totipotent cells can form all the cell types in a body, plus the extra-embryonic, or placental, cells. Embryonic cells within the first couple of cell divisions after fertilization are the only cells that are totipotent. Pluripotent cells can give rise to all of the cell types that make up the body; embryonic stem cells are considered pluripotent. Multipotent cells can develop into more than one cell type but are more limited than pluripotent cells; adult stem cells and cord blood stem cells are considered multipotent. Oligopotent cells are also called progenitor cells and can differentiate into a few cell types. Finally, unipotent cells, also called precursor cells, can give rise to only one type of differentiated cells. Choices C and D are incorrect since the sequence of decreasing potency is not followed.

✅ Q.8.2. **The correct answer is C.** The function of MHC I molecules is to display peptides of exogenous proteins to cytotoxic T lymphocytes (CTL) and not to T-helper cells. The choices A and D are incorrect. MHC class I antigens are found on the surface of every nucleated cell and are the predominant reason for rejection of allogeneic grafts.

✅ Q.8.3. **The correct answer is B.** Discovery of induced pluripotent stem cells (iPSCs) is considered a major scientific breakthrough, mainly because it overthrew a long-standing dogma that differentiation process cannot be reversed. Other choices are incorrect as it was previously known that stem cells exist and can yield different types of specialized cells, and such differentiation process can occur in both human and animal subjects.

✅ Q.8.4. **The correct answer is A.** It is important to disperse and re-seed stem cells at lower densities, because when colonies start to merge, stem cells can uncontrollably differentiate. Choice B is incorrect. Cell overgrowth can lead to media becoming more acidic but not more alkaline. Choice C is incorrect. When colonies start to merge, a process called contact inhibition will inhibit cell proliferation rates and not increase it. Choice D is incorrect. Contamination can result from cultivation media or cell culturing processes being unsterile and not cell overgrowth.

✅ Q.8.5. **The correct answer is B.** Immunoglobulins that develop against the epitope of interest are called primary antibodies. Choice A is incorrect as there are three (i.e., recognition, attack, and memory) and not five phases of acquired immune rejection. Choice C is incorrect. Acute rejection is mediated by cytotoxic T *lymphocytes*. There are no cytotoxic T *macrophages*. Choice D is incorrect. Transplantation of undifferentiated pluripotent stem cells can often lead to teratomas. The latter remains one of the major drawbacks of stem cell therapies.

Chapter 9 Answers

✓ Q.9.1.   **The correct answer is A.** The role of DNAse during the decellularization process is to chop nucleic acids released from damaged cells so it can be then removed by detergent solutions. Choice B is incorrect. Lipase or detergents are used to dissolve lipids, not DNAse. Choice C is incorrect. The main ingredient of ECM structure is a network of collagen fibers, which can be digested with the help of collagenase. Since the goal of decellularization is to remove cells while leaving ECM structure maximally intact, collagenase is not used during the decellularization process. Choice D is incorrect. DNAase does not hydrolyze peptide bonds.

✓ Q.9.2.   **The correct answer D.** TWEEN-20 is a nonionic detergent. Other choices are incorrect as Triton X-100 is nonionic, SDS is ionic, and CHAPS is zwitterionic detergent.

✓ Q.9.3.   **The correct answer is C, 30–40 micron.** This pore size corresponds to the dimension of an average isolated cell. It was shown experimentally to yield the best recellularization outcomes. Pores of lesser sizes (choices A and B) make hard for seeded cells to fit in, while pores >100 microns (choice D) are too large.

✓ Q.9.4.   **The correct answer is C.** The least important outcome to document after completion of tissue decellularization is the degree of its stiffness. The other three choices (A, B, D) are incorrect. It is very important to remove cellular content to ensure that DCT is free from remnants of detergents and salts and to maintain the integrity of the remaining ECM. Otherwise implanted DCT can cause immune rejection and inflammation of the host.

✓ Q.9.5.   **The correct answer is B.** Heating above 50 °C will lead to protein denaturation. The presence of denatured proteins, in turn, can cause an inflammatory response from the host. Other choices, such as freezing and altering osmolarity, are widely used to help release cells from the tissues.

Chapter 10 Answers

✓ Q.10.1.   **The correct answer is D.** Unless temperature drops below freezing, cooling has a minimal adverse effect on cell viability post-printing. Factors listed as choices A–C can all cause irreversible damage to cells during the 3D printing process.

✓ Q.10.2.   **The correct answer is C.** Linear dimensions of individual isolated cells vary within a relatively narrow range (20–30 micron); therefore, if cell concentrations within bioink are similar, the type of cells within it will have minimal impact on the amount of extruded material. The other three choices will have an immediate direct effect on the amount of extruded ink and the dimensions of the extruded lines. Assuming all other printing parameters are the same, more viscous ink will be pushed out less, yielding more narrow lines. The higher velocity of needle tip movement will yield more narrow lines, again assuming all other printing parameters are the same. Larger gauge number corresponds to a narrower needle and lesser diameter of extruded lines.

✅ Q.10.3. **The correct answer is A.** The surrounding mold made from agarose has to be broken to release the cast sphere (in case of more flexible silicone molds it might be possible to release the sphere without breaking it). Cube and cylinder can be released from the agarose mold without breaking it by lifting these shapes up. To be released without breaking the mold, the cone shape has to be initially positioned with its flat surface facing upward.

✅ Q.10.4. **The correct answer is D.** Unless it is a specifically designed material, magnesium ions do not affect bioink solidification. Choices listed in A–C are commonly used to aid solidification of wide varieties of bioinks. Calcium ions are used to aid the polymerization of alginate and collagen bioinks, while cold temperatures will help to solidify gelatin-based bioinks. The UV exposure is used to cause polymerization of GelMA-based bioinks using so-called UV initiators, such as Irgacure or LAP (lithium phenyl-2,4,6-trimeth ylbenzoylphosphinate).

✅ Q.10.5. **The correct answer is D.** Changes in osmolarity do not lead to GelMA solidification. In contrast, all other choices (pairs listed under A–C) suggest either chemical or physical way to gel respective bioink.

Chapter 11 Answers

✅ Q.11.1. **The correct answer is B.** Tissue-engineered vessel is likely to have a cylindrical shape. Therefore, the use of flow bioreactor that provides nutrients to the lumen of cannulated vessel should be the best choice. Perfusion fluid can also be made to flow outside the vessel to help feeding the cells closer to the outside wall. Choices A and D are incorrect as they will provide less natural flow pattern and are more suitable for smaller and more compact pieces of tissue. Choice C is more suitable for tissue-engineered bone or cartilage.

✅ Q.11.2. **The correct answer is A.** Electrical stimulation and cyclical stretch are the main physical factors shown to promote maturation of engineered heart constructs. Choice B is incorrect. It can be applicable to many metabolically demanding cells and not necessarily lead to cardiac construct maturation. Choice C is incorrect since such conditions are not physiological, and periodic changes in extracellular calcium concentrations are not observed in vivo. Choice D is incorrect since mechanical tissue compression is more suitable to help the maturation of engineered bone.

✓ Q.11.3. **The correct answer is D.** Choice D is correct. Mimicking mechanical tissue compression has been shown to help the maturation of engineered bone. Choice A is incorrect since electrical stimulation does not mimic the physiological environment of bone development and is more suitable to promote maturation of engineered heart constructs. Choice B is incorrect. Daily media change is often used to culture many metabolically demanding tissues but not the bone. Choice C is incorrect since such conditions are not physiological, and periodic changes in extracellular calcium concentrations are not observed in vivo.

✓ Q.11.4. **The correct answer is D.** Tissue vascularization strategies include all listed approaches including in vivo vascularization of sequential layers of engineered tissue, creation of small artificial channels using microfluidics, and incorporation of angiogenic factors stimulating endothelial cell proliferation.

✓ Q.11.5. **The correct answer is C.** General bioreactor design must enable a user to regulate temperature, oxygen delivery, and rate of media perfusion as well as sterilize bioreactor for repeated use. The ability to observe cells while they are forming tissue is a highly beneficial feature available only in selected types of bioreactors.

Chapter 12 Answers

✓ Q.12.1. **The correct answer is D.** The main goal of the poster presentation is to outline results and conclusions in a visually clear way. Other choices are incorrect. Choices A and C are applicable to a peer-review article. Choice B is usually reserved for non-scientific presentations as too much of visual effects will only distract the audience from the take-home message.

✓ Q.12.2. **The correct answer is A.** The recommended number of slides for oral presentation is about one slide per minute. Choice B will lead to an extremely slow-paced presentation, while in case of choices C and D content of the presentation will be too rushed.

✓ Q.12.3. **The correct answer is B.** The easiest way to use the same figures across different formats (poster, paper, or slide presentation) is to make graphics black while using a white background. Other choices are less practical. In the case of choice A, figures might look good during a slideshow or poster presentation, but they have to be redone for publication purposes. Choice C is incorrect as it will lead to poor readability. Choice D is incorrect since both colors are customarily used for warning signs, and their hue and intensity will distract viewers from the main message.

✅ Q.12.4. **The correct answer is B**. Conference halls often get crowded, and so viewers should be able to see the main content of the poster from approximately 1.5–2 m since this average distance between poster rows is about 3–4 m. Choice A is incorrect. A small amount of humor is very much appreciated during the oral presentation, and it makes the audience more receptive to the presenter's main message. Choices C and D are incorrect. Oral presentation needs to highlight the key findings of the study. The Acknowledgment section should include not only the funding sources but also the names of the people who assisted authors with their studies.

✅ Q.12.5. **The correct answer is C.** Most space in a typical science poster has to be given to the Results section, while all other sections have to be succinct.

Printed in the United States
By Bookmasters